Communications
in Computer and Information

T0238869

Commenced Publication in 2007
Founding and Former Series Editors:
Phoebe Chen, Alfredo Cuzzocrea, Xiaoyong Du, Orhun Kara, Ting Liu,
Dominik Ślęzak, and Xiaokang Yang

More information about this series at http://www.springer.com/series/7899

Qingfeng Chen · Jia Wu
Shichao Zhang · Changan Yuan
Lynn Batten · Gang Li (Eds.)

Applications and Techniques in Information Security

9th International Conference, ATIS 2018
Nanning, China, November 9–11, 2018
Proceedings

 Springer

Editors
Qingfeng Chen
Guangxi University
Nanning, China

Jia Wu
Macquarie University
Sydney, NSW, Australia

Shichao Zhang
Guangxi Normal University
Nanning, China

Changan Yuan
Guangxi College of Education
Nanning, China

Lynn Batten ⓘ
Deakin University
Geelong, VIC, Australia

Gang Li ⓘ
Deakin University
Geelong, VIC, Australia

ISSN 1865-0929   ISSN 1865-0937 (electronic)
Communications in Computer and Information Science
ISBN 978-981-13-2906-7   ISBN 978-981-13-2907-4 (eBook)
https://doi.org/10.1007/978-981-13-2907-4

Library of Congress Control Number: 2018957411

This Springer imprint is published by the registered company Springer Nature Singapore Pte Ltd.
The registered company address is: 152 Beach Road, #21-01/04 Gateway East, Singapore 189721, Singapore

Preface

The International Conference on Applications and Techniques in Information Security (ATIS) has been held annually since 2010. This year, the ninth in the ATIS series was held at Guangxi University, Nanning, China, during November 9–11, 2018. ATIS focuses on all aspects on techniques and applications in information security research, and provides a valuable connection between the theoretical and the implementation communities and attracts participants from industry and academia.

The selection process this year was competitive; we received 59 submissions, which reflects the recognition of and interest in this conference. Each submitted paper was reviewed by three members of the Program Committee. Following this independent review, there were discussions among reviewers and chairs. A total of 19 papers were selected as full papers, with the acceptance rate of 32%. Moreover, we were honored to have two prestigious scholars giving keynote speeches at the conference: Prof. Shuo Bai (Chinaledger Technical Committee, China) and Prof. Shui Yu (University of Technology Sydney, Australia).

We would like to thank everyone who participated in the development of the ATIS 2018 program. In particular, we would give special thanks to the Program Committee, for their diligence and concern for the quality of the program and as well as for their detailed feedback to the authors. The conference also relied on the efforts of the ATIS the 2018 Organizing Committee. In particular, we thank Prof. Ling Song, Ms. Haiqiong Luo, Mr. Wei Lan, and Mr. Zhixian Liu for the general administrative issues, the registration process, and maintaining the conference website.

Finally and most importantly, we thank all the authors, who are the primary reason why ATIS 2018 was so exciting and was a premier forum for presentation and discussion of innovative ideas, research results, applications, and experience from around the world as well as highlight activities in the related areas. Because of your great work, ATIS 2018 was a great success.

September 2018   Qingfeng Chen
                                                                        Jia Wu

Organization

ATIS 2018 was organized by the School of Computer, Electronic and Information, Guangxi University, Nanning, China.

ATIS Steering Committee

Lynn Batten (Chair)	Deakin University, Australia
Heejo Lee	Korea University, South Korea
Gang Li (Secretary)	Deakin University, Australia
Jiqiang Liu	Beijing Jiaotong University, China
Tsutomu Matsumoto	Yokohama National University, Japan
Wenjia Niu	Beijing Jiaotong University, China
Yuliang Zheng	University of Alabama at Birmingham, USA

ATIS 2018 Organizing Committee

General Co-chairs

Shichao Zhang	Guangxi Normal University, China
Changan Yuan	Guangxi College of Education, China
Lynn Batten	Deakin University, Australia

Program Committee Co-chairs

Qingfeng Chen	Guangxi University, China
Jia Wu	Macquarie University, Australia

Organization Co-chairs

Cheng Zhong	Guangxi University, China
Taosheng Li	Guangxi University, China

Awards Co-chairs

Xiaohui Cheng	Guilin University of Technology, China
Xianxian Li	Guangxi Normal University, China

Liaison Co-chairs

Jialiang Hai	Qinzhou University, China
Zhong Tang	Guangxi Medical University, China

Publicity Co-chairs

Xiaolan Xie	Guilin University of Technology, China
Ling Song	Guangxi University, China

Publication Co-chairs

Srikanth Prabhu	Manipal Institute of Technology, India
Ziqi Yan	Beijing Jiaotong University, China

Organizing Committee

Daofeng Li	Guangxi University, China
Baohua Huang	Guangxi University, China
Cheng Luo	Guangxi University, China
Jia Xu	Guangxi University, China
Wei Lan	Guangxi University, China
Zhixian Liu	Guangxi University, China
Akshay M. J.	Manipal Institute of Technology, India

Web Masters

Qiong Wu	Xi'an Shiyou University, China
Ye Lei	Xi'an Shiyou University, China

Program Committee

Edilson Arenas	Central Queensland University, Australia
Leijla Batina	Radboud University, The Netherlands
Guoyong Cai	Guilin University of Electronic Technology, China
Yanan Cao	Chinese Academy of Science, China
Liang Chang	University of Manchester, UK
Morshed Choudhury	Deakin University, Australia
Bernard Colbert	Deakin University, Australia
Rohan DeSilva	Central Queensland University, Australia
Xuejie Ding	Chinese Academy of Sciences, China
William Guo	Central Queensland University, Australia
Xin Han	Xi'an Shiyou University, China
Jin B. Hong	University of Canterbury, New Zealand
Rafiqul Islam	Charles Sturt University, Australia
Meena Jha	Central Queensland University, Australia
Jianlong Tan	Chinese Academy of Sciences, China
Dawei Jin	Zhongnan University of Economics and Law, China
Dong Seong Kim	University of Canterbury, New Zealand
Kwangjo Kim	KAIST, South Korea
Jie Kong	Xi'an Shiyou University, China
Chi-Sung Laih	National Chung Kung University, Taiwan
Qian Li	Chinese Academy of Sciences, China
Shu Li	Chinese Academy of Sciences, China
Yufeng Lin	Central Queensland University, Australia
Qingyun Liu	Chinese Academy of Sciences, China
Shaowu Liu	University of Technology, Sydney, Australia
Wei Ma	Chinese Academy of Sciences, China

Tsutomu Matsumoto   Yokohama National University, Korea
Meng Ren   Xi'an Shiyou University, China
Eiji Okamoto   University of Tsukuba, Japan
Katsuyuki Okeya   Hitachi Systems Development Laboratory, Japan
Lei Pan   Deakin University, Australia
Udaya Parampalli   University of Melbourne, Australia
Biplob Ray   Central Queensland University, Australia
Wei Ren   China University of Geosciences, China
Rei Safavi-Naini   University of Calgary, Canada
Jinqiao Shi   Chinese Academy of Sciences, China
Lisa Soon   Central Queensland University, Australia
Dirk Thatmann   Technische University Berlin, Germany
Brijesh Verma   Central Queensland University, Australia
Steve Versteeg   CA, Australia
Hongtao Wang   Chinese Academy of Sciences, China
Jinlong Wang   Qingdao University of Technology, China
Qian Wang   Zhongnan University of Economics and Law, China
Xiaofeng Wang   Siemens Research, China
Marilyn Wells   Central Queensland University, Australia
Gang Xiong   Chinese Academy of Sciences, China
Ping Xiong   Zhongnan University of Economics and Law, China
Yuemei Xu   Beijing Foreign Studies University, China
Rui Xue   Chinese Academy of Sciences, China
Fei Yan   Wuhan University, China
Ziqi Yan   Beijing Jiaotong University, China
Feng Yi   Chinese Academy of Sciences, China
Xun Yi   RMIT University, Australia
Chengde Zhang   Southwest Jiaotong University, China
Lefeng Zhang   Zhongnan University of Economics and Law, China
Yuan Zhang   Nanjing University, China
Yongbin Zhou   Chinese Academy of Sciences, China
Feng Zhu   Chinese Academy of Sciences, China
Liehuang Zhu   Beijing Institute of Technology, China
Tianqing Zhu   Wuhan University of Technology, China
Tingshao Zhu   Chinese Academy of Sciences, China
Yujia Zhu   Chinese Academy of Sciences, China

Sponsoring Institutions

Guangxi University, China
Guangxi Normal University, China
Guilin University of Technology, China
Qinzhou University, China
Deakin University, Australia
Macquarie University, Australia

Contents

Knowledge Discovery

Applications

Information Security

Optimization of Location Information Hiding Algorithm for Anti Attack

Bin Wang[1,2], Rong-yang Zhao[3(✉)], Guo-yin Zhang[1],
and Jia-Hai Liang[3]

[1] College of Computer Science and Technology, Harbin Engineering University,
Harbin 150001, China
[2] College of Information and Electronic Technology, Jiamusi University,
Jiamusi 154007, China
[3] College of Electronics and Information Engineering, Qinzhou University,
Qinzhou 535011, China
36782349@qq.com

Abstract. The wireless communication network, often automatically positioning information about a user's location, location privacy for the effective protection of the user, put forward a kind of position information hiding method based on Improved Genetic Algorithm for large data position information of a piecewise linear encoding processing, feature extraction of location information based on association rules, using vector quantization method for information fusion the data after encoding hidden design, improved genetic algorithm with location information encryption key construction, location information encryption and encoding processing, to achieve information hiding location optimization. The simulation results show that the design method of hiding location information, the location information of the good encryption performance, strong anti attack capability, improve the privacy protection performance of location information, and the encryption algorithm for the hidden cost is low, high real-time encoding of information hiding.

Keywords: Improved genetic algorithm · Privacy protection
Location information · Encryption · Coding

1 Introduction

With the development of the Internet and mobile communication technologies, people are engaged in information fusion processing and big data storage under the cloud computing environment through mobile Internet terminal devices, providing convenience for people's work and life. However, the location information in the data information processing platform is vulnerable to hackers attacks and easily stolen, which results in information security risks. Therefore, it is necessary to encrypt and hide the location information. The research on location-related information hiding methods has received great attention [1].

A location information hiding method based on improved genetic algorithm quantization coding is proposed in this paper. Firstly, piecewise linear coding is performed on the position information big data, the association rule characteristic quantity

Q. Chen et al. (Eds.): ATIS 2018, CCIS 950, pp. 3–11, 2018.
https://doi.org/10.1007/978-981-13-2907-4_1

of position information is extracted, and vector quantization fusion method is used to hide the encoded information. Then, the location information is encrypted and quantized by the improved genetic algorithm, and the position information hiding optimization is realized. Finally, simulation experiments were performed.

2 Preprocessing and Piecewise Linear Coding the Position Information Bit Sequence

2.1 Preprocessing the Position Information Bit Sequence

In order to effectively hide the position information, the position information bit sequence needs to be preprocessed. We use Setup model, Extract-Partial-Key model and Identity Authentication model to build a cryptosystem [7]. Assuming that the position information bit sequences $a_0 a_1 a_2 \cdots a_n$ and $b_0 b_1 b_2 \cdots b_n$ are n-dimensional binary sequences controlled by the position information scheduling parameters p. G_1, G_2 are assumed to be random distribution models of order p. The source coding sequence of the location information is $s = \{s_i, i = 1 \ldots M \mid s_i \in S\}$. Where S represents a binary vector quantization function for randomly selected n−1 position sources $m_j \in Z_p^*(1 \leq j \leq n, j \neq k)$. Data characteristics were mine by adaptive piecewise linear equalization method. Which is expressed as $e : G_1 \times G_1 \rightarrow G_2$. Encrypted the ciphertext of multiple character sequences, we obtained the key set $e : G_1 \times G_1 \rightarrow G_2$ within the distribution interval of the position information bit sequence $p_0, \ldots, p_{l-1}$. In an encrypted cycle, we use the three retransmission method to reconstruct the ciphertext sequence, as shown in Fig. 1.

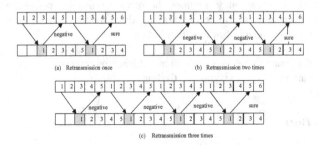

Fig. 1. Three retransmission mechanism of encrypted bit sequence

According to Fig. 1, the position information hidden mapping relationship can be obtained. A threshold cryptosystem is established using a Gaussian random partition method, and a key parameter needed for encryption is obtained, expressed as:

$$CT_{ID_i} = (C_1 = upk_{i1}^r,$$
$$C_2 = upk_{i2}^r;$$
$$C_3 = me(g_1, g_2)^r e(g_1, g^{u_i(H_1(ID_i, upk_i) - H_1(g, g_1, g_2, g_3, h))})^r,$$
$$C_4 = Te(g_1, g_2)^r e(g_1, g^{u_i(H_1(ID_i, upk_i) - H_1(g, g_1, g_2, g_3, h))})^r,$$
$$C_5 = 1) \tag{1}$$

Randomly selected the symbols in the location information distribution link layer $h \in G_1 \backslash \{1_{G_1}\}$, $h_0 \in G_2 \backslash \{1_{G_2}\}$, $\zeta_1, \zeta_2 \in Z_p^*$, and $\{1_{G_1}\}$, $\{1_{G_2}\}$ performed piecewise mapping. Statistical threshold P-value:

$$P\text{ - value} = 2[1 - \varphi(S_{obs})]$$
$$= 2\left(1 - \frac{1}{\sqrt{2\pi}} \int_{-\infty}^{S_{obs}} e^{-u^2/2} du\right)$$
$$= \frac{2}{\sqrt{2\pi}} \int_{S_{obs}}^{+\infty} e^{-u^2/2} du$$
$$= \frac{2}{\sqrt{\pi}} \int_{\frac{S_{obs}}{\sqrt{2}}}^{+\infty} e^{-t^2} dt \tag{2}$$
$$= erfc\left(\frac{S_{obs}}{\sqrt{2}}\right)$$

If the threshold P − value ≥ 0.01, the group functions of the location information distribution are G1 and G2.

2.2 Piecewise Linear Coding

It is assumed that the position posting message M = {group ID ‖ message ID ‖ payload ‖ timestamp}. Under the mobile Internet platform, we randomly select location information encryption parameters $\alpha, \beta, r_\alpha, r_\beta, r_{x_i}, r_{\delta_1}, r_{\delta_2} \in Z_p^*$, calculating $T_1 = u^\alpha$, $T_2 = v^\beta$, $T_3 = A_i h^{\alpha+\beta}$, $\delta_1 = x_i \alpha$, $\delta_2 = x_i \beta$; calculating $R_1 = u^{r_\alpha}$, $R_2 = v^{r_\beta}$, $R_3 = e(T_3, g_2)^{r_{x_i}} e(h, w)^{-r_\alpha - r_\beta} e(h, g_2)^{-r_{\delta_1} - r_{\delta_2}}$, $R_4 = T_1^{r_{x_i}} u^{-r_{\delta_1}}$, $R_5 = T_2^{r_{x_i}} v^{-r_{\delta_2}}$; And counting the binomial sum of the encrypted bit sequence X $S_n = x_1 + x_2 + \ldots + x_n$, Initializing public parameters $param = (G_1, G_2, G_T, p, \psi, e, H)$, we obtained a position informa-tion hidden group public key $gpk = (g_1, g_2, u, v, w, h, h_0, h_1, h_2)$, computing $c = H(M, T_1, T_2, T_3, R_1, R_2, R_3, R_4, R_5) \in Z_p^*$; calculating $S_\alpha = r_\alpha + c\alpha$, $S_\beta = r_\beta + c\beta$, $S_{x_i} = r_{x_i} + cx_i$, $S_{\delta_1} = r_{\delta_1} + c\delta_1$, $S_{\delta_2} = r_{\delta_2} + c\delta_2$; The privacy preserving bit sequence of location information is characteristics reorganized, and the reorganization result $\sigma = (T_1, T_2, T_3, R_1, R_2, R_3, R_4, R_5, S_\alpha, S_\beta, S_{x_i}, S_{\delta_1}, S_{\delta_2})$ are filled to the signature field to be encrypted in segments. The interval lengths of the plaintext sequence distribution are:

$$I^i = \sigma f^{-1}(x)(I^{i+1}) \tag{3}$$

$$size(I^i) = P_i size(I^{i+1}) \tag{4}$$

Then the location information in the block distribution is as follows:

$$\begin{aligned} size(I^1) &= \Pi_{i=1}^{M} P(s_i \in S) \\ &= \Pi_{n=1}^{N} (P_n)^{card\{s_i | s_i = S_n\}} \\ &= \Pi_{n=1}^{N} (P_n)^{P_n M} \end{aligned} \tag{5}$$

According to the corresponding cumulative probability, the position information genetic evolution frequency detection result is obtained as:

$$-\log_2(size(I^1)) = -\sum_{n=1}^{N} P_n M \log_2(P_n) = M \bullet H \tag{6}$$

The algebra of genetic evolution is an integer $\Pi_{i,b} = \chi_{i,b}^{\Pi} - \delta_{i,b}^{\Pi} (1 \le i \le \mu)$. The self-adaptive iteration and parameter optimization control were used for the number of location information hidden layers and improved the depth of data hiding [9].

3   Improving Genetic Algorithm and Optimizing Location Information Hiding Algorithm

3.1   Improving Genetic Algorithm

A location information hiding method based on improved genetic algorithm and quantization coding is proposed in this paper. Taking a three-layer neural network as an example, after multiple iterations, when $\lim_{n \to \infty} P\left(\frac{S_n}{\sqrt{n}} \le z\right) = \varphi(z) = \frac{1}{\sqrt{2\pi}} \int_{-\infty}^{z} e^{-u^2/2} du$, the chemokine of genetic evolution is $q(x_k^i / x_{k-1}^i)$. The new genetic evolution scheduling set is $\{\tilde{x}_k^i\}_{i=1}^{N}$. Using this as the input weight, the global optimal scheduling of position information hiding is performed. Through the scientific preparation of the gene, the optimal inertia weight for position hiding is solved, and the test sequence set for the position hiding inertia weight is obtained:

$$X = \{x_1, x_2, \cdots, x_n\} \subset R^s \tag{7}$$

The data set contains n samples. Dividing the set of limited data into C classes, and $1 < c < n$. When the selected individuals are cross variant, the individual distribution vector with a vector point space of $i = 1, 2, \cdots, n$ is as follow:

$$x_i = (x_{i1}, x_{i2}, \cdots, x_{is})^T \tag{8}$$

After rounds of selection, location information was arithmetical coded, and the fuzzy clustering center matrix was obtained:

$$V = \{v_{ij} | i = 1, 2, \cdots, c, j = 1, 2, \cdots, s\} \tag{9}$$

Where V_i is the ith vector in the clustering characteristic space. Establishing the initial population, designing a fitness function of a population, and the obtained partition matrix of the position information encoding was as follows:

$$U = \{\mu_{ik} | i = 1, 2, \cdots, c, k = 1, 2, \cdots, n\} \tag{10}$$

Setting the variable to be solved as Q, where Xi is the solution in the space of the variable Q, and the genetic evolution objective function of position information hiding is defined as:

$$J_m(U, V) = \sum_{k=1}^{n} \sum_{i=1}^{c} \mu_{ik}^m (d_{ik})^2 \tag{11}$$

In the formula, m is the diversity agglomeration index, $(d_{ik})^2$ is the measure distance between each of corresponding Xi evolution samples x_k and V_i. A piecewise linear encoding process was performed on the position information big data, and the association rule characteristic number of the position information is extracted as:

$$\tilde{w}_k^i = \tilde{w}_{k-1}^i \frac{p(z_k / \tilde{x}_k^i) p(\tilde{x}_k^i / x_{k-1}^i)}{q(\tilde{x}_k^i / x_{k-1}^i)} \tag{12}$$

p, q represent the similarity and relevance of the position data, respectively. Calculating the corresponding price of each chromosome cost(popt(tree)) [10]. Through the linear equalization process, an adaptive weight coefficient with hidden position information is obtained:

$$\tilde{\tilde{w}}_k^i = \tilde{w}_k^i / \sum_{i=1}^{N} w_k^i \tag{13}$$

Thus, the location information hiding association rules are extracted.

3.2 Encryption and Quantization Coding for Location Information

Performing vector quantization encoding on the optimal individual chromosome particle group $\{x_0^i, i = 1, 2, \ldots, N\}$, for each Xi, the mapping relation is as follows:

$$l_i(k) = (1 - \rho) l_i(k - 1) + \gamma f(x_i(k)) \tag{14}$$

And Xi is fitness function, ρ represents the probability that the ith particle moves to the jth particle in its neighbor set at k times. The characteristic space recombination is performed on the privacy protection bit sequence of the location information to obtain the optimal solution function of the shift key within the corresponding cumulative probability interval:

$$P_{ij}(k) = \frac{(l_j(k) - l_i(k))\eta_{ij}(k)}{\sum\limits_{j \in N_i(k)} (l_j(k) - l_i(k))\eta_{ij}(k)} \tag{15}$$

The position of chromosome i in D-dimensional space can be expressed as $X_i = (x_{i1}, x_{i2}, \cdots x_{iD})$, The best position experienced by the chromosome after the i-th genetic evolution is recorded as $P_i = (p_{i1}, p_{i2}, \cdots p_{iD})$. Introducing an identity-based authentication and encryption system, and the encrypted output of location information was obtained:

$$\begin{cases} \mu_0 = (P_{01} + P_{02})/2 \\ \mu_1 = (P_{11} + P_{12})/2 \\ \mu_2 = (P_{21} + P_{22})/2 \\ \quad \cdots \cdots \\ \mu_L = (P_{L1} + P_{L2})/2 \end{cases} \tag{16}$$

$$\mu = 3.57 + \frac{1}{L}\sum_{i=0}^{L} \mu_i \tag{17}$$

The location privacy protection data encryption key is constructed as shown in Eq. (18), where H1 is the encryption master key:

$$\begin{aligned} &H_1 : F_q \times G_1 \rightarrow \{0,1\}^n, H_2 : \{0,1\}^* \rightarrow G \\ &H_3 : \{0,1\}^* \times \{0,1\}^* \rightarrow F_q, H_4 : \{0,1\}^n \rightarrow \{0,1\}^n \end{aligned} \tag{18}$$

Thus, the location information hiding encryption algorithm is optimized.

4 Simulation Experiment Analysis

In order to test the application performance of this method in implementing location information hiding and privacy protection, simulation experiments were conducted. In the experiment, the position information under different block lengths was subjected to arithmetic coding and encryption processing to realize information hiding. The length of the plaintext block for each set of location data was 12000. The number of data encryption iterations was set to 100. The measure coefficient of forward error correction coding was 0.24. The source code had a size of 12. The security parameter was 0.982. According to the normal distribution characteristics, the probability distribution coefficient of location information coding is 0.763. The bit sequence distribution of the link transmission layer of the location information distribution is:

10010100101001010100100101010010111010100101001010010100010010101001
010100101001001010010010100101010010010101001010010101010101001010

The method in this paper was compared with the method described in the document [5] and document [6] under the same experimental conditions, and the comparison results of the execution time when encrypting different location information are shown in Fig. 2. The analysis shows that the algorithm in this paper has less overhead of encryption time, higher execution efficiency. And it can encode and hide information timely.

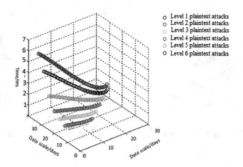

Fig. 2. Analysis of information encryption time of the proposed method

Figure 3 is a comparison of information hiding encryption performance of different methods under the attacks of known plaintext and self-selected plaintext. As can be seen from Fig. 3, under the attack of known plaintext, the method in this paper is completely unaffected, and the performance of concealing the encrypted protection location information is still 100%. While, under the same conditions, the methods described in the document are less affected, but the overall performance can still reach over 99%. Under the strong adaptive selection plaintext attacks, the hidden protection performance of the method proposed in this paper is still up to 100%. While, the hidden protection performance of the method described in document [5] is 86.9%. And the hidden protection performance of the method described in document [6] is 75.3%. This shows that the method described in this paper has stronger resistance to attacks.

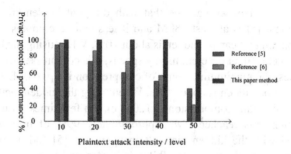

Fig. 3. Comparison of information hiding encryption performance under different attacks

In order to verify the effectiveness and feasibility of the improved hidden method, the experiment takes the location information hiding time as an indicator and a comparative analysis was conducted. The results are shown in Figs. 4, 5 and 6.

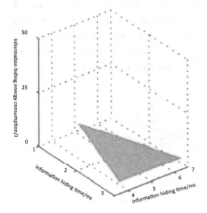

Fig. 4. Hidden method in this paper

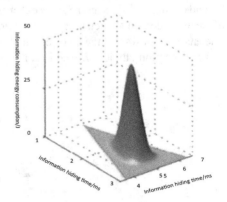

Fig. 5. Document [5] hiding method

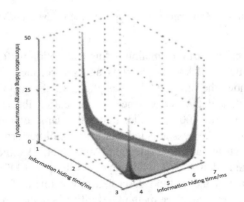

Fig. 6. Document [6] hiding method

From Figs. 4, 5 and 6, we can see that with different hidden speeds, the compression in document [5] is at least 280M and it less can be compressed to 0M. The amount of information that can be hidden is about 20%; When using the hiding method described in document [6], at different hiding speeds, it is compressed to 300M at least. In only a few cases, the maximum amount of compression is up to 0M. The amount of information that can be hidden is about 40%; When using the hidden method described in this paper, the minimum compression is 120M only a few times. In most cases, the compression amount has reached 0M. Approximately 99% of information can be hidden. Compared with the hidden methods in document [5] and document [6], the hidden effect is more significant, the feasibility is higher, and it has certain advantages.

5   Conclusion

Location information is easily attacked and stolen by hackers in the data information processing platform, resulting in leakage of user's location information privacy and potential information security risks. In order to improve the privacy protection capabilities of mobile network users, it is necessary to encrypt and hide the location information. This paper proposed a method of position information hiding based on improved genetic algorithm quantization coding. The ciphertext sequence reconstruction is performed using three retransmission methods within one encryption cycle. Boneh group signature was used to segmented coding the location information. Using genetic evolutionary algorithm for position information hiding coding and encryption algorithm optimization. Which has Improved the hiding ability of position information. The research shows that this method has a strong ability to hide location information, and has better privacy protection performance under attack, and is of great application value.

Acknowledgment. This research was supported by the National Science Foundation of Heilongjiang Province (Grant No. F2015022), the University Nursing Program for Young Scholars with Creative Talents in Heilongjiang Province (project number: UNPYSCT-2017149), The Key Laboratory for Advanced Technology To Internet of Things (IOT2017A03), Guangxi Young And Mid Aged Teachers' Basic Ability Promotion Project (2017KY0803), all support is gratefully acknowledged.

References

1. Sun, S.: A new information hiding method based on improved BPCS steganography. Adv. Multimed. **15**(3), 1–7 (2015)
2. Liu, H., Wang, X., Chen, Y., et al.: Optimization lighting layout based on gene density improved genetic algorithm for indoor visible light communications. Opt. Commun. **390**, 76–81 (2017)
3. Chen, X., Sun, H., Tobe, Y., Zhou, Z., Sun, X.: Coverless information hiding method based on the Chinese mathematical expression. In: Huang, Z., Sun, X., Luo, J., Wang, J. (eds.) ICCCS 2015. LNCS, vol. 9483, pp. 133–143. Springer, Cham (2015). https://doi.org/10.1007/978-3-319-27051-7_12
4. Gurung, S., Choudhury, K.P., Parmar, A., et al.: Multiple information hiding using cubical approach on random grids. J. Food Sci. **7**(11), 54–63 (2015)
5. Lu, J.-H., Long, C.-P.: Optical wireless transmission under the hidden image simulation reversible information hiding method. Comput. Simul. **34**(5), 201–204 (2017)
6. Zhang, X.-M., Yin, X.: Speech information hiding approach in wavelet domain based on chaotic sequence. J. Syst. Simul. **19**(9), 2113–2117 (2007)
7. Wang, J.H., Wang, J.L., Wang, D.M., et al.: Neural network location based on weight optimization with genetic algorithm under the condition of less information. IEICE Trans. Commun. **99**(11), 2323–2331 (2016)
8. Zou, Y., Zhang, Y.: Improved multi-objective genetic algorithm based on parallel hybrid evolutionary theory. Environ. Sci. Technol. **34**(3), 133–134 (2015)
9. Basu, A., Nandy, K., Banerjee, A., et al.: On the implementation of IP protection using biometrics based information hiding and firewall. Int. J. Electron. **103**(2), 177–194 (2016)
10. Mahboubi, H., Moezzi, K., Aghdam, A.G., et al.: Distributed deployment algorithms for improved coverage in a network of wireless mobile sensors. IEEE Trans. Industr. Inf. **10**(1), 163–174 (2014)

Multi-user Order Preserving Encryption Scheme in Cloud Computing Environments

Taoshen Li$^{(\boxtimes)}$, Xiong Zhou, and Ruwei Wang

Guangxi University, Nanning 530004, People's Republic of China
tshli@gxu.edu.cn, dongz0202@foxmail.com,
ruweih@126.com

Abstract. Cloud computing, a modern application that is very commonly utilized, has garnered considerable interest from researchers and developers. Data privacy security is the most urgent issue related to cloud computing. Studies have shown that order-preserving encryption (OPE) technology is an effective way of securing privacy in the cloud. This study proposes a multi-user order-preserving encryption scheme in cloud environments (MUOPE). The scheme locates the multi-user problem to the most common many-to-many model according to security requirements. Before encryption plaintext, the original plaintext is randomly divided into successive intervals having different lengths, and the plaintexts are encrypted by encryption function. Trusted key generation center is introduced to generate user key and a corresponding auxiliary key for each user, and the encrypted ciphertext is re-encrypted by proxy re-encryption. The re-encrypted ciphertexts allow user to decrypt by using own private key. Security analysis and experimental have verified the safety and effectiveness of the MUOPE scheme.

Keywords: Order preserving encryption · Cloud computing · Multi-user
Local sensitive hashing · Ciphertext Re-encryption

1 Introduction

Cloud computing, as a remote service model, has been widely used. But security is its biggest challenge [1]. The risk of data leakage is particularly concerning when data is being evaluated in the cloud. Security mapping studies shows that the primary areas of weakness in the cloud are data protection (30.29%) and identity management (20.14%) [2].

Typically, user privacy can be protected effectively through encryption. The common solution to solve the problem of data privacy security in the cloud environment is to encrypt the data in advance before storing it to the cloud server and then decrypt it by the user when necessary. However, most encryption programs do not support ciphertext evaluations, such as the retrieval or operation of the ciphertexts [3, 4], so that traditional encryption technology is unable to meet the security needs of cloud environments.

There is an urgent need to encrypt data in databases to protect sensitive data. But, encrypted data cannot be efficiently queried because the data has to be decrypted. To resolve this problem, many systems use order-preserving encryption scheme (OPE). However, the existing OPE schemes cannot hide the original data distribution and are

© Springer Nature Singapore Pte Ltd. 2018
Q. Chen et al. (Eds.): ATIS 2018, CCIS 950, pp. 12–28, 2018.
https://doi.org/10.1007/978-981-13-2907-4_2

susceptible to statistical attacks, and the comparisons operation can be directly performed on encrypted data using OPE schemes. In addition, many OPE schemes are only applicable to single-user scenarios, or can be used in cases where multiple users are assumed to be completely trusted. These application scenarios are not consistent with real cloud computing applications. Therefore, the study of the OPE scheme that can hide the original data distribution and support multi-user retrieval is an urgent problem of ciphertext retrieval in the cloud environment.

For the above scenario, this paper proposes a multi-user order-preserving encryption scheme in cloud environments (MUOPE), which it combines ciphertext sorting and multi-user integration and is more suitable for cloud environments. The scheme not only retains the performance and security advantages of the OPE scheme, but also resists statistical attacks and collusion attacks. The following sections demonstrate the MUOPE scheme's capability to resolve privacy issues in cloud environments.

2 Related Works

Order-preserving encryption is an effective encryption scheme where the sort order of ciphertexts matches the sort order of the corresponding plaintexts, which allows databases and other applications to process queries involving order over encrypted data efficiently. OPE scheme ensures that comparing encrypted data returns the same result than comparing the original data. This permits to order encrypted data without the need of decryption. Studies have shown that OPE technology is an effective way of securing privacy in the cloud, and it can quickly retrieves the results that satisfy the condition without revealing any information other than the order. However, there are still some security issues in the OPE scheme. Some different improved OPE schemes have been proposed, and the scheme is now a rather hot topic in the cryptography field.

In order to reduce the computational complexity, Liu and Wang [5, 6] use a simplified linear function and noise to encrypt the plaintext, which plaintext encryption process is simple and high-performance, but it is easy to suffer from chosen-plaintext attack (CPA) and has low security. In [7], Boldyreva et al. present a modular order-preserving encryption (MOPE) which the scheme improved the security of OPE in a sense, as it does not leak any information about plaintext location. Krendelev et al. [8] propose two alternative OPE schemes based on arithmetic coding and sequence of matrices. Although the two schemes have higher security, the process of generating non-degenerate matrices in the scheme need to take a lot of computing time, and the time complexity will increase linearly with the complexity of the matrix exponential operation. In [9], Martinez et al. propose an order preserving symmetric encryption scheme whose encryption function is recursively constructed. At first, the scheme uses a binary search to determine the point's position, and takes the coordinates of the point on the diagonal as the encryption key. And then, the ciphertext value is calculated by a linear function. Fang et al. [10] present a new relaxed order preserving sub-matrices model by considering the linearity relaxation. To achieve the security standard, the scheme first divides the plaintext into multiple bucket sections, and then further divides the buckets by the pre-order tree, so that the number of data contained in each bucket is kept within a certain range.

Popa et al. [11] first present an order-preserving scheme that achieves ideal security, and Reddy et al. [12] also propose a new dynamic order-preserving encryption scheme which adopts the mutable OPE scheme that attains precise security. The two schemes use a technique called variable ciphertexts, meaning that over time, the ciphertexts for a small number of plaintext values change. But, both of these schemes require frequent interaction between the client and the server, and need the database server to maintain extra information and realize comparison or range query by user defined functions, which takes a lot of time and is not suitable in the actual cloud environment. In [13], Ahmadian et al. apply order preserving encryption scheme for the sensitive fields of a data record, and proposed encryption solution can supported various relational operations on encrypted data in hybrid cloud environments and balanced security concerns with efficiency for some applications. But the scheme required multiple computational operations when calculating hyper-geometric probabilities, and the computational load is large, so it was not suitable for large data sets. In [14], Liu et al. [10] propose an OPE scheme based on random linear mapping, which randomly divides the data in the original plaintext space into several subintervals, and then encrypts them with linear function and noise through. Since the interval is randomly divided, the same value may be divided into different intervals, and the obtained ciphertext value is different. In [15], Huang et al. design a randomized data structure-random tree, and construct an order-preserving encryption algorithm based on random tree (OPEART). The OPEART algorithm realizes the encryption of data by randomness, and supports relational calculations ($>$, $=$, etc.) on encrypted data. Security analysis and performance evaluation show that OPEART is IND-DNCPA while achieving the goal of relational calculations on encrypted cloud data efficiently. However, the ciphertext data encrypted by the OPEART algorithm maintains the same data distribution state as the original plaintext data, so the original data distribution cannot be well hidden.

Although the above OPE schemes can perform encryption, decryption and data retrieval efficiently, it has hidden dangers in security and is easy to suffer from attacks. Moreover, most of the existing OPE schemes can serve only a single recipient so they are not suitable for the multi-user application in cloud computing environment. Actual cloud environments have functions that may need to serve several users, such as access control, computing multi-user ciphertext, and multi-user sharing.

3 Problem Analysis and System Model

3.1 Application Problem Analysis

Consider the following scenario, employees of departments is limited access to company data based on their rating. The company trusts its cloud service provider to provide the resources necessary to complete its large-scale computing tasks; the data sender is independent of the company, and is not aware of the existence of other data senders. In order to meet the company's privacy needs, each data sender must set appropriate access policies to encrypt data, and each employee must meet the certification requirements for the encrypted data to gain access. When employees of different departments upload encrypted data, only the employees meeting the access requirements of the encrypted data can compute it into ciphertext, and only those meeting the access requirements for the resulting ciphertext can access it.

According to the situation of the data owner and the authorized data user in the above application scenario, various typical scenarios can be abstracted into four application models: one-to-one, one-to-many, many-to-one, and many-to-many. Where, the one-to-one model belongs to the single-user class, and the other three models belong to the multi-user class.

Because the MUOPE scheme proposed in this paper is to study how to apply the ciphertext sorting algorithm to multi-user environment, the most common many-to-many model is selected as the application model of this paper. If the ciphertext sorting algorithm supports many-to-many application scenarios, it will naturally support one-to-many and many-to-one scenarios. Therefore, the multiple users in this paper represent the case of multiple data owners and multiple authorized data users.

Under many-to-many application scenarios, multiple data owners can upload encrypted files to the server, and can retrieve and decrypt these ciphertext files. Therefore, the securities of many-to-many models are:

(1) Any user can generate encrypted files, and data owners and other legitimate data users can retrieve and decrypt data.
(2) Except for data owners and authorized data consumers, the server cannot obtain search conditions and plaintext data.

3.2 System Model

Figure 1 models the privacy protection support scheme in a typical cloud computing environment, reflecting the interaction among user group (UG), cloud database server (CDS) and key generation center (KGC).

The interaction process of the system is as follows:

(1) The user in the UG first encrypts the file content $p_i(i \in [1, n], n \geq 1)$ by using OPE scheme with a key s, and gets the ciphertext data set $c_i(i \in [1, n], n \geq 1)$. And then, the user uploads c_i to the CDS through a secure channel.
(2) The KGC center generates a set of key pairs (pk_i, sk_i) for each user, and the user encrypts the keyword w by using the public key pk_i and uploads the keyword ciphertext to the CDS.
(3) When the user needs to use the system to perform operations such as querying, the request parameter ($para$) is firstly encrypted by using the symmetric key s to obtain the query ciphertext ($para_1$). And then the private key sk_i is used for re-encryption to obtain the query trapdoor E ($para_1$), and the E ($para_1$) and the calculation request $type$ are send to the CDS.
(4) After receiving the user's query trap, the CDS first verifies the user's query permission and decrypts the trapdoor E ($para_1$). Only the authorized user can correctly decrypt the query ciphertext. After the verification is passed, according to ($para_1$) and the calculation request, the CDS will return the ciphertext result E ($result$) that satisfies the condition.
(5) After receiving the E ($result$), the user decrypts the ciphertext by using the key s to get the corresponding plaintext.

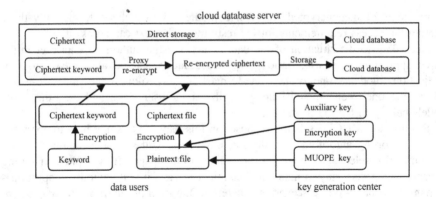

Fig. 1. Cloud computing system architecture supporting multi-user privacy protection

3.3   Security Analysis

As shown in Fig. 1, the system has three roles that are users, cloud service providers and key generation centers. Because the system's security analysis is based on some important features, it is necessary for following assumptions:

(1) The authorized users in the system are completely trusted. Only authorized users can perform queries, calculations, and so on. They do not start up collusion attack with the cloud service provider to steal private information and other internal attacks.
(2) The cloud service provider is honest and curious. For each query request, the server honestly executes the operation specified by the user and returns the correct result, but the result information is logged. In this way, the server may try to excavate out other information of the user from the recorded information.
(3) The KGC center is a fully trusted server that provides relevant key management and access control for the system without collusion attacks with cloud service providers.
(4) The revoked and unauthorized users are not trusted, and they may get to a collusion attack with other users for some purposes.

From the above discussion, we can find that there are two potential threats in the system: one is the cloud server, which will extract more private information from the storage and query process; the second is the revoked and unauthorized users, which they may get the ability to exceed their access rights by illegal means. Thence, we think that a multi-user OPE scheme should must meet the following three requirements:

(1) Data confidentiality. This is the most basic requirement of any encryption scheme. The user data (including plaintext and index) will not be stolen by illegal users and cloud servers. Even if the ciphertext is stolen by the enemy, it will not be decrypted.
(2) Anti-collusion attack. A collusion attack is an attack that tries to more privileges by merging private keys between users. The scheme should be able to withstand such attacks.

(3) Anti-statistical attacks. The cloud server may mine out more private information from the storage and query process, and obtain extra information through statistical analysis. This is an internal attack on cloud computing. In addition, the attacker can also find some ciphertext values whose distribution characteristics are not obvious from the plaintext data, and then start up other attacks.

4 First Section Our Scheme

4.1 Design Idea

This section will construct a multi-user order-preserving encryption scheme (MUOPE), which supports any relational operation on encrypted data, and supports multiple users and can resist collusion attacks. The design idea of the scheme is as follows:

(1) According to the security requirements of the application scenario, the multi-user problem is positioned as many-to-many model.
(2) Before encrypting the plaintext, the original plaintext space is first randomly divided into consecutive intervals with different lengths, then a larger range of expansion space is selected and divided into the same number of intervals, and then the original plaintext is mapped to the new expansion space by using the mapping function. The mapping functions can be different for different expansion space. Because the plaintext space is randomly divided, the same plaintext may be mapped to different expansion spaces, so that the distribution probability of the original data can be hidden.
(3) The value of the extended space is used as the input of the encryption algorithm and is encrypted by an encryption function. At first, the encryption function obtain an initial value range and an initial interval field. After each round of calculation, the value range of the ciphertext will be reduced, and the size of the interval field will be randomly changed. Finally, the ciphertext value is determined within a data field by random number. By using the random number, the encrypted data cannot be predicted, so that ensures a high security of the MUOPE scheme ensuring.
(4) The proxy re-encryption technology is used to re-encrypt the encrypted data by the KGC center. Other users can use their own private key to decrypt the re-encrypted ciphertext in order to get the first encrypted ciphertext. Since the first encryption uses the order-preserving encryption method, the decrypted ciphertext still guarantees the order of the ciphertext, and can be directly operated on the ciphertext.

4.2 Definitions of MUOPE Scheme

Definition 1: Let the domain of the original plaintext be $D \in [mMin, mMax]$ and $mMin$ and $mMax$ are the minimum and maximum values of the domain D respectively, the value range of the encryption algorithm be $V \in [vMin, vMax]$ and $vMin$ and $vMax$ are the minimum and maximum values of the encryption algorithm respectively, and the intermediate value domain be C is $\in [tMin, tMax]$ and $tMin$ and $tMax$ are the minimum and maximum values of the domain MS respectively. The Multi-user order-preserving encryption scheme MUOPE = (*Init, KeyGen, Encrypt, ReEncrypt, Trapdoor, Search, Decrypt, Revoke*) consists of the following five algorithms:

(1) The initialization algorithm *Init*: $\{tMin, tMax, v\} \leftarrow Init(mMin, mMax, m)$, where the v is intermediate value obtained by processing the original plaintext m. The task of data initialization algorithm *Init* is to divide the D and then map it to the new extended space C to ensure the flatness of the data.

(2) The key generation algorithm *KeyGen*: $\{vmin, vmax, interval, level, num\} \leftarrow KeyGen(|C|, dk)$, where the dk is the data encryption key. The algorithm *KeyGen* assigns dk generated by the KGC center to each authorized user, and ensures that the decryption keys obtained by the user are the same.

(3) The encryption algorithm *Encrypt*. For data $d_i \in D$, $c_i \leftarrow Encrypt$ ($vMin, vMax, interval, level, num, dk, d_i$), where $c_i \in V$.

(4) The re-encryption algorithm *ReEncrypt*. For the ciphertext $c \in V$. $v \cup \phi \leftarrow ReEncrypt$ ($vmin, vmax, interval, level, num, dk, c$), where ϕ represents no solution. After re-encrypting, the data set (u, C, I') is actually stored on the server.

(5) The trapdoor generation algorithm *Trapdoor*. When the user needs to send a retrieval request, the retrieval parameter p is first encrypted by the key dk to obtain the ciphertext p_1, and the p_1 is re-encrypted by the user key xu_1 to obtain the trapping gate p_2. The query information set (u, p_2) is then sent to the server.

(6) The search algorithm *Search*. When the CDS receives the user query request, it first determines where user the query request comes from, and then uses the auxiliary key xu_2 to decrypt the trapdoor p_2. If the decryption is invalid, the user u is an unauthorized user and the query request cannot be accepted. If the decryption is correct, the obtained ciphertext p_1 is matched with the keyword I', and the corresponding ciphertext result is sent to the user u.

(7) The decryption algorithm *Decrypt*. For the ciphertext $c_i \in V$. $v_i \cup \phi \leftarrow Decrypt$ ($vMin, vMax, interval, level, num, dk, c_i$), where ϕ represents no solution.

(8) The revoking algorithm *Revoke*. After the user u authority is revoked, the algorithm *Revoke* can be used to delete the auxiliary key xu_2 corresponding to u on the cloud server, and u will not be able to send the query request again.

4.3   The Construction of MUOPE Scheme

Let G and GT denote bilinear groups with prime order p, g be generators of group G, and $e: G \times G \to GT$ be bilinear maps. Let H_1 and H_2 be two collision-resistant hash functions, respectively $H_1: \{0, 1\}^* \to G$, $H_2: G_T \to \{0, 1\}^*$. Finally, let *interval* is the initial interval domain size, *level* is the number of layers in the processing interval domain, *num* is the number of domain segmentation. The detailed construction of the OPEMU scheme is given below.

Initialization Algorithm *Init*. As mentioned earlier, the task of data initialization algorithm *Init* is to divide the D and then map it to the new extended space C to ensure the flatness of the data. The plaintext space D is will be divided into a series of intervals D_i ($i = 1, 2, ..., m$). The division of plaintext space can effectively hide the distribution law of the original data. Also, C is also divided into equal numbers of intervals C_i ($i = 1, 2, ..., m$) to further hide the data distribution. The mapping function is implemented by the following linear expression:

$$y = l'_i + scale * (x - l'_i) \tag{1}$$

Where, $x \in D_i$ is an original plaintext data; l'_i ($i = 1, 2, ..., m$) is the minimum values of the MC_i; $scale$ is the calculation scale factor, and the calculation formula is as follows:

$$scale = (r'_i - l'_i)/(r_i - l_i) \tag{2}$$

Where, l'_i and r'_i ($i = 1, 2, ..., m$) are the minimum and maximum values of the C_i respectively; l_i and r_i ($i = 1, 2, ..., m$) are the minimum and maximum values of the D_i respectively.

By mapping the original plaintext space D into the extended space MC, it is possible to effectively hide the distribution of the original data and change the probability of data distribution.

Key Generation Algorithm *KeyGen*. At first, after inputting the security parameter 1^k, the trusted key generation center selects appropriate public parameter ($p, g, G, G_T, e, H_1, H_2$) and randomly generates a data encryption key dk, and assigns it to each authorized user. And then, the KGC center generates a user key xu_1 and a corresponding auxiliary key xu_2 for each user u. After calculating $k_1 \cdot k_2 \equiv k(\bmod q)$, ($x_{u1}, k_1$) is sent to the user u, and (u, xu_2, k_2) is sent to the cloud server as the auxiliary key of the user u.

Encryption Algorithm *Encrypt*. At first, the user u uses the algorithm *Encrypt* to the encrypt the plaintext file d by key dk, and gets the ciphertext $C = (d, dk)$. Then, using xu_1 as the key, the keyword w is encrypted to obtain the keyword ciphertext $I_w = (w, xu_1)$. The ciphertext tuple (u, C, I_w) is uploaded to the server. The main pseudo code of the encryption algorithm *Encrypt* is described as follows:

```
procedure Enc(vmin,v max, interval, level, num, dk, dᵢ)
    t← (vmax-vmin)/num;
    tt←interval;
    for i=0 to level
        splitList←dᵢ
        ranList←random.next();
        borderList←border;
    end for
    for i=0 to level
        vmin = vmin + t * splitList[i] + tt * borderList[i];
        if split[i]== borderList[i][0]
        vmax = vmin + t * ranList[i] + tt * (borderList[i][1] – borderList[i-1][1]);
        else  max = min + t * ranList[i];
        tt = t * (1-ranList[i]);
    end for
    cᵢ = vmin + (vmax -vmin) * random(dk+ dᵢ);
end procedure
```

For each keyword w, the algorithm *Encrypt* establishes an index $I_w = H_2(e(H_1(w), g)^{ik})$. Finally, ciphertext tuple (u, C, I_w) is uploaded to the server.

In the process of encrypting data, each round of calculation changes the size of the initial value range by the random changed size of the interval. After multiply calculations, the ciphertext value is finally determined by a random number.

Re-encryption Algorithm *ReEncrypt*. The re-encryption algorithm *ReEncrypt* is probabilistic. When the server receives the ciphertext tuple (u, C, I) from user u, the CDS first re-encrypts (u, C, I) with the auxiliary key xu_2 associated with u to obtain the data set (u, C, I'). The data set is the ciphertext data that is actually stored on the server.

Trapdoor Generation Algorithm *Trapdoo*. When the user needs to send a retrieval request and query the keyword w, the processing of the algorithm *Trapdoo* is as follows:

Step1: retrieve the parameter p and encrypt it by the key dk to obtain the ciphertext p_1: $p_1 \leftarrow$ Encrypt ($vMin$, $vMax$, $interval$, $level$, num, dk, w).
Step2: Re-encrypt p_1 by the user key xu_1 to obtain the trapping gate p_2. The operation process is as follows:

$$p_2 = H_2\left(e(H_1(\mathrm{E}(w), x_{u1}), g)^{ik}\right) \tag{3}$$

Step3: Send the query information set (u, p_2) to the server.

Search Algorithm *Search*. After receiving the query request of user u, the processing of Search algorithm is as follows:

Step1: Verify user permissions.

1.1 Using the auxiliary key x_{u2} as the decryption key of the trapdoor p_2, perform the following operation:

$$E(w) = H_1\left(e(H_2(p_2, x_{u2}), g)^{ik}\right) \tag{4}$$

1.2 If $E(w)$ is empty, the decryption is invalid, and the user u is an unauthorized user and the query request is not accepted;

1.3 If $E(w)$ is not empty, the decryption is correct, and the user u has the query permission and his query request is processed.

Step2: Match the ciphertext p_1 with the keyword $E(w)$, and query the file set C_i related to $E(w)$.

Step3: Return C_i to user u.

Decryption Algorithm *Decrypt*. When the user u receives the ciphertext, the algorithm *Decrypt* is used to decrypt the ciphertext by using key dk. After decrypting the ciphertext c_i, the intermediate value v_i can be obtained. The main pseudo code of the decryption algorithm *Decrypt* is described as follows:

> procedure Decrypt (vMin, vMax, interval, level, num, dk, c)
> $t \leftarrow (vMax - vMin)/num$;
> $tt \leftarrow interval$;
>   for i=0 to level
> $x = \left(c - vMin - tt * \sum_{j=1}^{i-1} random(dk + x_j)\right)\big/ d$;
> $r = random(dk + x)$;
> $vMin = vMin + x * t + tt * \sum_{j=1}^{i-1} random(dk + x_j)$;
>   if $x == x_i$ $vMax = vMin + t * r + tt * random(dk + x_j)$;
>   else $vMax = vMin + t * r$;
> $tt = t * (1 - r)$;
> $v = v + x$;
>   end for
> end procedure

Revoking algorithm *Revoke*. After the authority of the user u is revoked, the processing of the algorithm *Trapdoo* is as follows:

Step1: The auxiliary key x_{u2} corresponding to the u is deleted on the cloud server.

Step2: The user authority table maintained by the cloud server is updated.

After the user u is revoked, it is not able to send the query request again.

5 Security Analysis

Theorem 1: Anti-collusion attack. An encryption algorithm is safe under collusion attacks if it satisfies that the user can't get more permission by merging the private keys of other users.

Proof: Suppose the attribute private key of the user u_i is $SK_{u_i^j,k}$. After the attribute private key $SK_{u_i^j,k}$ of the user u_j is merged into the private key of the user u_i, we get the new attribute private key k, then:

$$
\begin{aligned}
SK_{u_i^j,k} &= (D, \{(D_i, D_i')\}_{\lambda_i \in I_{u_i^j,k}} \cup \{(D_j, D_j')\}_{\lambda_j \in I_{u_i^j,k}}) \\
&= (g^a g^{r_t^i}, \{g^{r_t^i} H(\lambda_i)^{r_i}, g^{r_t^i} g^{v_t^i}\}_{\lambda_i \in I_{u_i^j,k}} \cup \{g^{r_t^i} H(\lambda_j)^{r_j}, g^{r_t^j} g^{v_t^j}\}_{\lambda_j \in I_{u_i^j,k}})
\end{aligned}
\tag{5}
$$

Where, r_t^i and r_t^j are random numbers selected by users u_i and u_j respectively, and v_t^i and v_t^j are random numbers selected by the KGC center for different users. Suppose we want to decrypt the user u_i, $\forall \lambda_i \in I_{u_i^j,k}$, there are:

$$
SK_{u_i^j,k} = \frac{e(D_i, C_i)}{e(D_i', C_i')} = \frac{e(g^{r_u^i}, g^{q_i(0)})}{e(g^{v_t^i - v_t^j}, H(\lambda_i)^{q_i(0)})}
\tag{6}
$$

Because of $v_t^i \neq v_t^j$, the parameter s can't be calculated, and the user cannot obtain more permissions by merging the private keys of other users.

Obviously, the encryption algorithm in the MUOPE scheme is an encryption algorithm that conforms to Theorem 1, so it can resist collusion attacks.

Definition 2: An encryption algorithm E is safe under statistical attacks, if an attacker ξ initiates the attack under the following conditions:

(1) The distribution probabilities of the original data are not equal;
(2) Suppose $p(x \in D_1) > p(x \in D_2) > \ldots > p(x \in D_m)$. After multiple queries, the ciphertext distribution probability obtained by ξ is $p(x \in C_1)$, $p(x \in C_2)$, ..., $p(x \in C_m)$.
(3) Based on statistical analysis and the plaintext data distribution, the guess made by ξ is: $p(x \in C_1) > p(x \in C_2) > \ldots > p(x \in C_m)$.

Definition 3: Advantage probability of the ξ: $Adv(\xi) = |p(x \in C_i - p(x \in C_j)| = \varepsilon$, $(i, j \in [1, m]$ and $i \neq j)$. If ε is small enough, the encryption algorithm E is safe under statistical attacks.

Definition 4: If $D_i \in D$ is a interval and $x \in D_i$ is a number, then $p(x \in D_i) = |D_i|/L(D_i)$. Where $L(D_i)$ is the width of D_i, the $|D_i|$ is the element numbers of D_i.

Theorem 2: The MUOPE algorithm is safe under statistical attacks.

Proof: In the data initialization process, the MUOPE algorithm can hide the distribution of the original data and change the probability of data distribution through division operation and mapping operation. According to definition 2, $p(x \in D_i) = |D_i|/L(D_i)$. If there is no spatial expansion, the ciphertext distribution is the same as the plaintext distribution. Since $p(x \in D_1) > p(x \in D_2) > ... > p(x \in D_m)$, there are $p(x \in C_1) > p(x \in C_2) > ... > p(x \in C_m)$. After spatial expansion, The plaintext space D is divided into a series of intervals D_i ($i = 1, 2, ..., m$) with different sizes. Also, C is also divided into equal numbers of intervals C_i ($i = 1, 2, ..., m$), and the size of the C_i is determined according to the element numbers of D_i.

Assume that element numbers of any two sub-intervals D_i, D_j in the plaintext interval are s and t respectively, then the sizes of the mapping interval C_i, C_j are $s \cdot g$, $t \cdot g$ (g is the set unit distance). The ciphertext interval $x \in C_1$, $x \in C_2$, ..., $x \in C_m$ obtained by the calculation will exhibit a near linear distribution: $y = k \cdot x + b$. Then, for the elements in D_i, D_j, there are:

$$p(x \in D_i) = |D_i|/L(D_i) = s/(s \cdot g) = 1/g$$
$$p(x \in D_i) = |D_i|/L(D_i) = t/(t \cdot g) = 1/g$$

Similarly, $p(x \in C_1) \approx p(x \in C_2) \approx ... \approx p(x \in C_m)$ can be derived, that is $|p(x \in C_i - p(x \in C_j)| = \varepsilon$, $(i, j \in [1, m]$ and $i \neq j)$, where ε is arbitrarily small.

Therefore, this proves that the MUOPE algorithm is safe under statistical attacks.

In the encryption process, since the MUOPE scheme utilizes the random number feature and the location of the ciphertext value is determined by the random number after multiple calculations, the encrypted data cannot be predicted. Therefore, the MUOPE scheme has high security and confidentiality.

6   Experiment and Performance Analysis

In order to analyze the execution efficiency of the MUOPE scheme, our compared the MUOPE scheme with the OPEART scheme proposed by [15]. The experimental performance analysis includes encryption and decryption time, retrieval time, storage/communication load and et al.

6.1   Experimental Environment

The experiment was deployed on a cloud wing experimental platform based on Web, KVM and Swift. KVM is hardware-based full virtualization tool which mainly responsible for virtual machine creation, virtual memory allocation, VCPU register read and write, and VCPU operation. Swift is used for long time storage of permanent types of static data that can be retrieved, adjusted, and updated. Examples of data types that are best suited for storage are virtual machine images, image storage, mail storage, and archive backups.

The server cluster of the platform configured with two high-profile servers and 10 standard servers. The configuration of a high-profile server 32G memory, 24 core, 1T hard disk. The configuration of the standard server was 4G memory, 2 core, 500G hard disk. The operating environment is set on several PCs (1.5G memory, 3.0 GHZ CPU), which operating system is Windows XP SP3. The grid platform tool is Globus Toolkit 4.0 and JDK1.6.

6.2   Data Distribution Probability Experiment

Experiment 1 mainly analyzes and compares the data distribution probability after encryption by MUOPE and OPEART. The experiment selects the salary-employee data as shown in Fig. 2. The experimental results of the salary date distribution after encryption by two schemes are shown in Fig. 3.

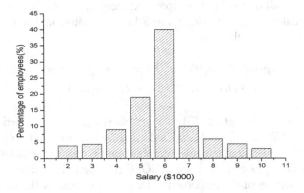

Fig. 2. Salary-employee data distribution of a company

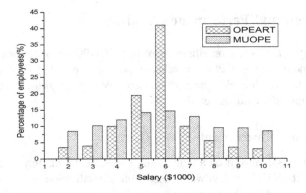

Fig. 3. Data probability distribution results after encryption

As can be seen from the experimental results, the data distribution probability after encryption by the OPEART scheme was almost the same as the original data distribution probability, and the data distribution probability after encryption by the MUOPE scheme was significantly different from the original data distribution probability, and the distribution probability of various types of salary was basically the same. When encrypting the data, the OPEART scheme does not linear segment and map the original data, so that the ciphertext and the plaintext exhibit the same data distribution probability. Before the data is encrypted, the MUOPE scheme first divides and maps the original data through the initialization algorithm, so that the data of each interval is uniformly distributed. Experimental results show that the MUOPE scheme can better hide the distribution probability of the original plaintext data, and the cloud server cannot obtain more private information and additional information from the storage and query process, and cannot carry out statistical attacks.

6.3   Algorithm's Efficiency Comparison Experiment

The purpose of experiment 2 is mainly to compare the encryption and decryption performance of MUOPE and OPEART. Set parameters *level* = 7, *num* = 99. Experimental testing data is shown in Figs. 4 and 5.

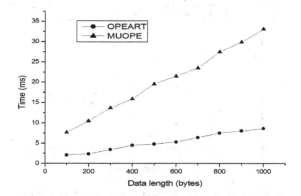

Fig. 4. Experiment comparison of encryption time

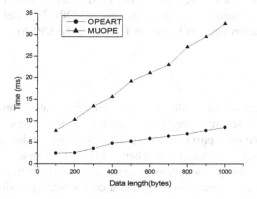

Fig. 5. Experiment comparison of decryption time

Compared to OPEART, the encryption and decryption time of MUOPE increased linearly. This is because the MUOPE scheme adds the initialization algorithm (*Init*) and the re-encryption algorithm (*ReEncrypt*) and the revocation algorithm (*Revoke*), so it takes more time than the OPEART time during the entire encryption and decryption process.

Experiment 3 mainly compares the encryption and search performance of MUOPE and OPEART. Set parameters *level* = 7, *num* = 99. Experimental testing data is shown in Fig. 6. The retrieval time of the MUOPE was slightly larger than the OPEART, and it increased linearly. This is because the MUOPE scheme needed to call the trapdoor generation algorithm to encrypt the user's retrieval parameter p to obtain the ciphertext p_1, and use the user key xu_1 to re-encrypt p_1 to obtain the trapping gate p_2, and then send the information group (u, p_2) obtained by the query to the server. These encryption and re-encryption processes costed a certain amount of time.

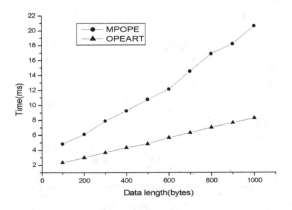

Fig. 6. Experiment comparison results of searching time

Based on the above experimental comparison results, although the encryption and decryption time and retrieval time of the MUOPE is larger than the OPEART, the difference is not significant. It is important that MUOPE is better able to hide the original data distribution and has higher security than the OPEART.

7   Conclusion

Aiming at the problem of using order preserving encryption algorithm under multi-user scenario, an order preserving encryption scheme (MUOPE) supporting multi-user is proposed, which combines proxy re-encryption technology, order-preserving encryption (OPE) technology and multi-users scheme. MUOPE scheme locates the multiple user problem to the most common many-to-many model, and the trusted key generation center is introduced to generate user key and a corresponding auxiliary key for each user. In the initialization phase, the scheme utilizes a linear expression to map the original plaintext space to the extended space, so it can hide the distribution of the original data and change the probability of data distribution. When encrypting, it uses the data of the extended space as the input value of the encryption algorithm, and the ciphertext's value is determined by introducing a random number. Finally, the proxy re-encryption mechanism is employed to re-encrypt the encrypted data. The re-encryption ciphertexts allow user to decrypt by using own private key. MUOPE scheme supports the multi-user environment and enables data owners to revoke data consumers dynamically and efficiently. Security analysis has proved that the MUOPE scheme can resist collusion attacks and statistical attacks on the basis of the OPE scheme. Experimental results also show that it can guarantee the privacy of data and further improve the performance of the sequence-preserving encryption scheme.

Acknowledgments. This work was partly supported by the National Natural Science Foundation of China (No. 61762010 and No. 61462007).

References

1. Modi, C., Patel, D., Borisaniya, B., Patel, A., Rajarajan, M.: A survey of intrusion detection techniques in Cloud. J. Netw. Comput. Appl. **36**(1), 42–57 (2013)
2. Belen, C.Z., Jose, L.F.A., Ambrosio, T.: Security in cloud computing: a mapping study. Comput. Sci. Inf. Syst. **12**(1), 161–184 (2014)
3. Wang, J.F., et al.: Efficient verifiable fuzzy keyword search over encrypted data in cloud computing. Comput. Sci. Inf. Syst. **10**(2), 667–684 (2013)
4. Huang, R.W., Gu, X.L., Yang, S., Zhou, W.: Study of privacy preserving framework for cloud storage. Comput. Sci. Inf. Syst. **8**(3), 801–819 (2011)
5. Liu, D.X., Wang, S.L.: Programmable order-preserving secure index for encrypted database query. In: 2012 IEEE Fifth International Conference on Cloud Computing, pp. 502–509. IEEE Computer Society, Honolulu (2012)
6. Liu, D.X., Wang, S.L.: Nonlinear order preserving index for encrypted database query in service cloud environments. Concurr. Comput.: Pract. Exp. **25**(13), 1967–1984 (2013)
7. Boldyreva, A., Chenette, N., O'Neill, A.: Order-preserving encryption revisited: improved security analysis and alternative solutions. In: Rogaway, P. (ed.) CRYPTO 2011. LNCS, vol. 6841, pp. 578–595. Springer, Heidelberg (2011). https://doi.org/10.1007/978-3-642-22792-9_33
8. Krendelev, S.F., Yakovlev, M., Usoltseva, M.: Order-preserving encryption schemes based on arithmetic coding and matrices. In: Proceedings the 2014 Federated Conference on Computer Science and Information Systems, pp. 891–899. Institute of Electrical and Electronics Engineers Inc., Warsaw (2014)
9. Martinez, S., Miret, J.M., Tomas, R., et al.: Securing databases by using diagonal-based order preserving symmetric encryption. Appl. Math. Inf. Sci. **8**(5), 2085–2094 (2014)
10. Fang, Q., Wilfred, N., Feng, J.L., et al.: Bucket order-preserving sub-matrices in gene expression data. IEEE Trans. Knowl. Data Eng. **24**(12), 2218–2231 (2012)
11. Popa, R.A., Li, F.H., Zeldovich, N.: An ideal-security protocol for order-preserving encoding. In: 2013 IEEE Symposium on IEEE Security and Privacy, pp. 1–15. Institute of Electrical and Electronics Engineers Inc., San Francisco (2013)
12. Reddy, K.S., Ramachandram, S.: A novel dynamic order-preserving encryption scheme. In: 2014 First International Conference on Networks and Software Computing, pp. 92–96. Institute of Electrical and Electronics Engineers Inc., Guntur (2014)

13. Ahmadian, M., Paya, A., Marinescu, D.C.: Security of applications involving multiple organizations and order preserving encryption in hybrid cloud environments. In: IEEE 28th International Parallel and Distributed Processing Symposium Workshops, pp. 894–903. IEEE Computer Society, Phoenix (2014)
14. Liu, Z.L., Chen, X.F., Yang, G.J., et al.: New order preserving encryption model for outsourced database in cloud environments. J. Netw. Comput. Appl. **59**, 198–207 (2014)
15. Huang, R.W., Gui, X.N., Chen, N.: An encryption algorithm supporting relational calculations in cloud computing. J. Softw. **26**(5), 1181–1195 (2015)

Dynamic Memory Allocation of Embedded Real-Time Operating System Based on TLSF

Xiaohui Cheng$^{(\boxtimes)}$ and Haodong Tang

Guangxi Key Laboratory of Embedded Technology and Intelligent System,
College of Information Science and Engineering,
Guilin University of Technology, Guilin, Guangxi, China
cxiaohui@glut.edu.cn

Abstract. In order to meet the requirements of rapid real-time and high memory utilization in the dynamic memory allocation of embedded real-time operating systems, this paper introduces memory block access attributes into the process of memory allocation and merge based on the TLSF algorithm and designs a TLSF-based embedded real-time operating system dynamic memory management mechanism. In the process of dynamic memory frequent application and merger will increase the longer time consumption and higher memory fragmentation rate, especially in the long-running embedded real-time operating system is more obvious. This paper uses the bitmap and the two-level segregated list data structure of the TLSF algorithm to quickly locate free memory blocks. This way can improve the search rate of free memory blocks. And by using the statistical data such as system dynamic memory allocation, release, reapply interval and duration, we can acquire the access attributes of memory blocks of different sizes. These access attributes contribute to selecting different memory allocation and merge strategies. The experimental results on μC/OS-II operating system show that the improved embedded dynamic memory management algorithm in this paper can effectively improve the memory allocation efficiency while ensuring a low fragmentation rate.

Keywords: TLSF algorithm · Embedded system · Dynamic memory
Access attributes

1 Introduction

With the development of computer technology in the direction of intelligence and miniaturization, embedded real-time operating systems play an increasingly important role, and have been widely used in aerospace, military, transportation, medical and other research fields. The biggest difference between embedded real-time operating systems and other non-real-time operating systems is that they need to work continuously for a very long period of time. This time is usually in months, years or even never stop. This requires that the resource utilization of the embedded real-time operating system is higher than that of the general system to ensure that new application requests can be responded when the system resources are not exhausted, especially memory resources.

© Springer Nature Singapore Pte Ltd. 2018
Q. Chen et al. (Eds.): ATIS 2018, CCIS 950, pp. 29–39, 2018.
https://doi.org/10.1007/978-981-13-2907-4_3

Memory management is divided into static memory allocation and dynamic memory allocation. Static memory is pre-allocated before the program runs, while dynamic memory is dynamically applied and released during the running of the program. In order to better control the operation of the program and the real-time requirements of the program, dynamic memory is generally adopted in the embedded system. After several decades of development, the dynamic memory has achieved good development results, mainly including sequential search, partner system, Bitmap matching and TLSF [1] and other classic dynamic memory allocation algorithms. The advantages and disadvantages of these algorithms have become an important factor affecting the performance of embedded systems. The high real-time operating system requirements for memory resources also bring some technical problems, mainly focusing on how to improve memory utilization, shorten memory allocation time and estimate the worst allocation time. The dynamic memory algorithm based on the TLSF algorithm is a hotspot in the field of embedded memory research. It mainly includes different initialization levels of memory blocks of different sizes [2], adjustment of memory cutting threshold [3, 4], and different ways to handle different sizes of memory [5]. On the basis of the literature [6], Wang et al. [7] improved the dynamic memory management of different memory block lifetimes and proposed a strategy of directly using short-lived memory blocks without merging, which causes to a large shortening in memory allocation time. Jiang et al. [8] studied the role of fuzzy threshold merging algorithm in the memory merge process from the application of video on demand system and also had a good effect.

Starting from the embedded real-time operating system, this paper aims at shorten the memory block allocation time and improve the TLSF dynamic memory allocation algorithm. The memory block access attribute is introduced to determine the memory block types of different sizes, and the allocation strategy of directly assigning different memory blocks to the table and dividing the large memory blocks is adopted. When recycling, a strategy combining no-merge and immediate merge is adopted. The test results on μC/OS-II system show that the improved dynamic memory allocation algorithm has faster memory allocation efficiency under the premise of ensuring lower memory fragmentation rate.

2   Related Work

Since the two-level Segregated Fit (TLSF) algorithm was proposed by Masmano et al. [1] in 2004, it has received the attention of embedded real-time operating system researchers because of its special data structure, simple search process and limited time complexity, which makes it become a research hotspot of the embedded real-time operating system dynamic memory allocation algorithm. The core principle of the TLSF algorithm will be briefly introduced from the aspects of the TLSF data structure and the process of finding free memory blocks.

2.1 TLSF Data Structure

The two most important concepts of TLSF are "two levels" and "separate fit", which link different sizes of free memory blocks by means of a two-level segregated list array. The first level divides the free block into 2n (n = 4, 5, ..., 31), which is defined as FLI (First-level Segregated Fit), and the second level linearly segments the first-level memory block according to user requirements, it is defined as SLI (Second-level Segregated Fit). At the same time, a related bitmap is created for each segregated list array to mark which segregated list contains free blocks (there are free blocks when the bitmap is 1 and no free blocks when 0). As shown in Fig. 1 below [1].

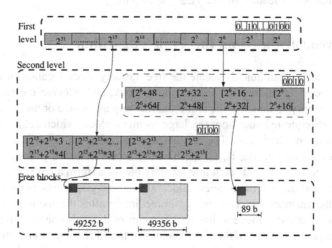

Fig. 1. TLSF data structure

2.2 TLSF Free Memory Block Search Process

As can be seen from the data structure described in Sect. 2.1 above, the process of finding the free memory block by the TLSF algorithm: when the application applies for a size memory block, the memory block size is first calculated by the following formulas (1) and (2) respectively. The first segregated list index F and the second segregated list index S.

$$F = \lfloor log_2(size) \rfloor \tag{1}$$

$$S = (size - 2^F) \frac{2^{sli}}{2^F} \tag{2}$$

Where SLI is the number of intervals in which the second segregated list is divided. Taking Fig. 1 as an example, the value of SLI is 4. If there is a free block at the position of the first segregated list index F, the free block is found according to the value of the second segregated list index S and assigned to the application. If there is no free block available for allocation in the position of the index of the first segregated list index F, then it is searched whether there is a free block at the index position of F+1, and if there has free memory blocks, this block will be allocated, if there is no, the search continues until the free block is found or the system is found. The report could not find a memory block that could be allocated.

2.3 TLSF Time Complexity

The process of TLSF searching for free memory blocks by bitmap and segregated free list data structure makes the average time of the algorithm constant, so the time complexity is O(1). Even in the worst case, the memory allocation time is predictable, which is an important feature of the TLSF algorithm.

3 Improvement Based on the TLSF Algorithm

The TLSF algorithm can quickly locate the free memory block location of the system during the process of finding the free memory block, and allocate the free memory block in the memory suitable for the requested memory block size or the memory block suitable for the requested size from the large memory block, which achieved a higher memory allocation efficiency. The TLSF algorithm has certain disadvantages in memory block recycling, that is, all the memory blocks released immediately in the systems is merged with the adjacent free memory blocks to form a larger memory block for the next allocation, and the frequent allocation and merge process make it takes a lot of time for the memory block to merge immediately after the memory is released, which is not conducive to the real-time requirements of the real-time system to a certain extent. In this paper, in view of the shortcomings of the TLSF algorithm in the memory allocation and merging process, a first-level segregated list bitmap is added to indicate the access type of different memory block sizes in the system memory allocation and release process. The determination of the type of the memory block is determined by the number of times the memory block is requested to be released, the time span of re-access, and the length of memory occupation, thereby determining the strategy that should be adopted when the memory block of this type is allocated and recycled.

3.1 Data Structure Definition

In the system, add a task-level data structure "memory access statistics table" and a system-level data structure "memory type bitmap", wherein the memory access statistics table MA_table has five attribute fields, the memory block size, the number of application, the number of releases, the interval of revisiting, and the occupied memory time used are as shown in Table 1 below.

Table 1. Memory Access Statistics Table MA_table

Memory size	Number of applications	Number of releases	Interval of revisiting	Occupied memory time

The memory type bitmap MT is consistent with the length of the first-level segregated list FLI in the TLSF data structure, and the value of each bit represents the type of the memory block of the class size, which is used to determine the allocation and recovery strategy of the memory blocks of different sizes. Based on the TLSF data structure in Fig. 1, a first level memory block type bitmap MT is added to represent each memory interval range to determine the allocation manner of each memory block size. A 0 on this bit indicates the merge policy immediately after partitioning and releasing from the large memory block at the time of allocation. A 1 on the bit indicates that the memory block of this size is directly acquired in the memory pool and is immediately reclaimed into the memory pool after being used and do not merge, just wait for the next memory application to be directly assigned. Set all bits of the MT are 0 or the lowest two bits are 1 in the initial state of the system. Since the memory blocks applied by the application are basically concentrated in a certain range and most of them are small memory blocks [9, 10], in the process of constant memory application and release, by obtaining the number of applications of different size memory blocks, the number of releases, the time interval of revisiting, the length of memory usage and the value of the bitmap in the MT can be changed in real time to adapt to different Application memory application requirements. After adding the memory type bitmap MT, the system can represents different size memory block segregated list data structure as shown in Fig. 2.

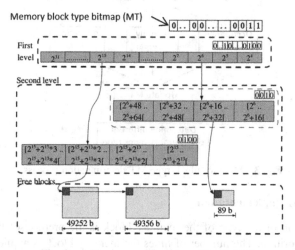

Fig. 2. Add memory block type bitmap MT

3.2 Memory Block Allocation and Merging

During the running of the system, the memory blocks that are frequently allocated and released by the system, the memory blocks that have been applied for a long time interval and the memory block information that takes a long time are obtained. According to the access statistics of different size memory blocks, the type bitmap MT representing different size memory blocks is continuously modified. In the process of the memory allocating, when the corresponding bitmap of the different size memory blocks in the MT is 1, the memory block of the size is directly obtained from the memory pool. When the value of the bitmap is 0, the size of the bitmap is the memory block is partitioned from the large memory block by way of bitmap and free list query. In the process of memory recycling, the system need to check the type of the memory block of the class size. If the corresponding bitmap of the memory block of the class size is 1, the merge operation is not performed, and the memory block is directly put into the memory pool. If the corresponding bitmap of the size memory block in the MT is 0, the memory block is merged with the adjacent free memory block according to the built-in merge strategy of the TLSF algorithm, and then the merged memory block is inserted into the corresponding free memory block segregated list. The process of determining the access type bitmap MT of different size memory blocks is as shown in Fig. 3.

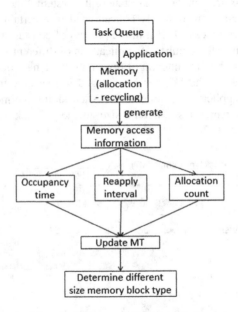

Fig. 3. Memory block type MT determination process

3.3 Algorithm Implementation

The determination of the type of the memory block is the premise of the implementation of the algorithm. The number of times the memory block is applied in the most recently set period of time is A, the number of times of being released is R, the time interval of being applied again is I, and the duration of memory occupation is T. Use these four parameters to determine the type of memory block. Defining S to present the difference between A and R, that is

$$S = A - R \tag{3}$$

When A is larger and S is closer to zero, it indicates that the memory block is frequently applied and released, and thresholds A' and S' are defined to determine whether the memory block of the this size is frequently applied and released, when A > A′ and S < S′, it is determined that the corresponding bitmap of the memory block of the class size in the MT is 1, otherwise it continues to determine the time interval I that is applied again. The threshold I′ is defined to determine whether I can be directly allocated. When I < I′, the bitmap corresponding to the size of the memory block in the MT is 1, otherwise the memory occupation duration T is continued to determine. The threshold T′ is defined to determine whether T can be directly allocated. When T > T′, the bitmap of the memory block of the class size in the MT is 1, otherwise the memory block of the class size is 0. The values of the maps in the MT change continuously with the constant memory application and release process during the system operation. The memory allocation and recycle process at any time changes according to the current value in the MT to adapt to the embedded real-time operating system. The application requires different memory blocks for different sizes. The improved memory allocation flow chart is shown in Fig. 4.

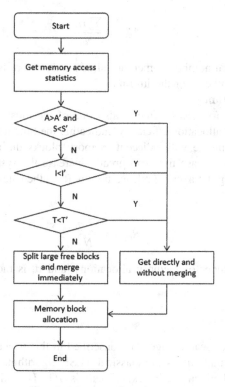

Fig. 4. Memory allocation process

4   Experiment and Result Analysis

4.1   Performance Test

In order to verify the performance of the embedded real-time operating system dynamic memory allocation described in this paper, it can be tested from the memory allocation time and the memory release time respectively. The system memory allocation efficiency is measured by the average memory allocation time and the average memory release time.

(1) **Average distribution time**

The average memory allocation time is one of the most commonly used criteria for verifying memory allocation efficiency. It represents the average time spent in the memory allocation process. The shorter the average allocation time, the stronger the real-time performance of the allocation algorithm in system memory allocation. Defining the average memory allocation time is A_{ave}, and the calculation formula is as follows (4).

$$A_{ave} = \frac{\sum rt_i}{N} \tag{4}$$

While N is the total number of memory allocations, and rt_i is the time it takes to allocate memory blocks for the ith time.

(2) **Average release time**

The average memory release time is also one of the most common criteria for verifying memory allocation efficiency, indicating the time it takes to insert into a memory pool or merge with adjacent memory blocks during memory release. A shorter memory release time can greatly increase the system's real-time performance. Defining the average release time is R_{ave}, the calculation formula is as follows (5).

$$R_{ave} = \frac{\sum rt_i}{N} \tag{5}$$

N is the total number of memory allocations, and rt_i is the ith memory block release time.

4.2   Analysis of Results

The dynamic memory allocation algorithm described in this paper is tested separately under the same test conditions as the classical TLSF algorithm. The size of the test memory partition pool given in the experiment is 1G. The allocation and release operations of memory blocks of different sizes are randomly generated in the five intervals [0, 16], [0, 32], [0, 64], [0, 128], [0, 256], [0, 512] respectively, and each group is tested 100 times in average to obtain the memory allocation time and the average memory release time. The average allocating time comparison chart is shown in Fig. 5 below and the average releasing time comparison chart is shown in Fig. 6 below.

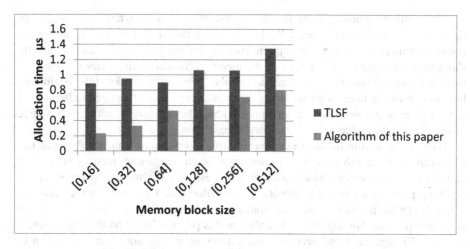

Fig. 5. Comparison of average memory allocation time

When the memory interval of the application is [0, 16] or [0, 32], the value of the MT bitmap is 1, so the memory block allocation speed of the algorithm is significantly lower than the TLSF algorithm. When the memory application interval is getting larger, the system needs to modify the MT value according to the frequency of the previous application and release of the memory, the time interval of the application again, and the time length of the memory, and then compare the value with the set threshold to modify the value of the MT. It is faster to apply for the same range of memory.

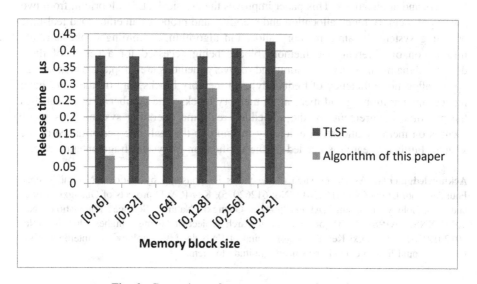

Fig. 6. Comparison of average memory release time

The minimum memory block size of the system is set to 16 bytes. Therefore, when applying for the memory interval [0, 16], the algorithm does not need to merge the released memory block with the adjacent free block during the merge, thereby greatly shortening the average release time of the memory. Similar to the memory allocation time, the system needs to merge a certain number of memory blocks at the beginning, but as the memory interval grows larger, many memory blocks that need to be merged can be directly put into the memory pool without being merged with adjacent free blocks. This reduces the average memory release time.

It can be seen from the above comparison results and analysis that in the same test environment, the improved dynamic memory allocation algorithm reduces the cutting operation in the memory block allocation process and the merge operation in the recovery process, so that the average of the allocation time and release time are reduced. The fragmentation rate is generated due to frequent cutting during the memory allocation process. The algorithm described in this paper is based on the improvement of the TLSF algorithm and performs less cutting, so it does not cause the system to generate higher fragmentation rate.

5   Conclusion

The research on the dynamic memory allocation mechanism of embedded real-time operating system focuses on three main aspects: speeding up allocating efficiency, reducing the number of merges and improving memory utilization. The optimization of distribution efficiency and the reduction of fragmentation rate can effectively guarantee the real-time response of the system and meet the requirements. Memory requirements for users and applications. This paper improves the classical TLSF algorithm from two aspects of memory block allocation and merging, and proposes an embedded real-time operating system dynamic memory allocation algorithm. According to the attribute information of different size memory blocks being accessed for a period of time, different dynamic memory allocation and recovery methods are adopted, which speeds up the allocation efficiency of frequently used memory blocks, and on the other hand reduces the partitioning and merging of memory blocks. The number of times satisfies the real-time requirements of the embedded real-time operating system. This paper focuses on memory allocation efficiency to improve the real-time performance of the system. Further research is needed on the verification of system fragmentation rate.

Acknowledgments. As the research of the thesis is sponsored by National Natural Science Foundation of China (No: 61662017, No: 61262075), Key R & D projects of Guangxi Science and Technology Program (AB17195042), Guangxi Natural Science Foundation (No: 2017GXNSFAA198223), Major scientific research project of Guangxi higher education (No: 201201ZD012), Guangxi Key Laboratory Fund of Embedded Technology and Intelligent System, we would like to extend our sincere gratitude to them.

References

1. Masmano, M., et al.: TLSF: a new dynamic memory allocator for real-time systems. In: Euromicro Conference on Real-time Systems, vol. 16, pp. 79–86 (2004)
2. Sun, Y., Wu, W., Zheng, J., Zhao, M., Li, B.: Design of an embedded real-time system dynamic memory manager. Small Comput. Syst. **35**(5), 1106–1110 (2014)
3. Song, M., Li, S.: A new embedded dynamic memory allocation algorithm. J. Comput. Appl. **37**(S2), 244–247, 254 (2017)
4. Liu, L., Zhu, Q., He, Z.: Analysis and improvement of FreeTROS memory management scheme. Comput. Eng. Appl. **52**(13), 76–80 (2016)
5. Shen, F., Zhang, Y., Lin, W.: Design of dynamic memory management algorithm in embedded real-time system. Comput. Mod. **7**, 103–107 (2015)
6. Ramakrishna, M., et al.: Smart dynamic memory allocator for embedded systems. In: International Symposium on Computer and Information, pp. 1–6 (2008)
7. Wang, X., Qiu, Y., Mu, F., Leng, Y.: Research on new dynamic memory management mechanism of embedded system. Microelectron. Comput. **34**(8), 66–69 (2017)
8. Jiang, Y., Zeng, X., Sun, P., Zhu, X.: Fuzzy threshold combined memory management algorithm for real-time embedded multimedia system. J. Xidian Univ. (Nat. Sci. Ed.) **39**(5), 174–180 (2012)
9. Berger, E.D., Zorn, B.G., McKinley, K.S.: Reconsidering custom memory allocation. SIGPLAN Not. **37**(11), 1–12 (2002)
10. Lu, X., Shuai, J., Wu, Q.: A new memory manager for object-oriented programs. Comput. Eng. **38**(9), 21–23 (2012)

Enhanced Authentication and Key Agreement Mechanism Using PKI

Krishna Prakasha[1], Pratheeksha Gowda[1], Vasundhara Acharya[2]([envelope]),
Balachandra Muniyal[1], and Mayank Khandelwal[3]

[1] Department of Information and Communication Technology,
Manipal Institute of Technology, MAHE, Manipal, India
{kkp.prakash,bala.chandra}@manipal.edu, pratheekshagowda21@gmail.com
[2] Department of Computer Science and Engineering,
Manipal Institute of Technology, MAHE, Manipal, India
vasundhara.acharya@manipal.edu
[3] Aalto University, Espoo, Finland
mayank.khandelwal@aalto.fi

Abstract. Entity Authentication and Key Agreement (AKA), is a critical cryptographic problem in wireless communication, where a mutual entity authentication plays a vital role in the establishment of the secure and authentic connection. The paper proposes an efficient authenticated key agreement scheme and increases the speed of authentication process more securely. The signaling overhead is minimized by creating the validity token at home agent of mobile equipment, which indicates if the certificate of the foreign agent is valid or invalid. An efficient way for the implementation of an enhanced version of the protocol is proposed. NTRU algorithm is applied to encrypt and decrypt the messages. NTRU algorithm is one of the efficient asymmetric key lattice-based cryptographic algorithm. NTRU has been proved to be the fastest and secure encryption algorithm. AES is used for symmetric key encryption. The result demonstrates that the proposed method is efficient.

Keywords: AKA · NTRU · OCSP · PKI · WPKI
Wireless communication

1 Introduction

In defiance of many advantages in the increase of mobile communication or wireless technology, the threat to security is also increasing. The balancing of growth in demand to provide network security services with the limitations of the devices is a significant challenge. The sustained demand for mobile devices causes the telecommunications network to attain the focus on its performance. A customer can make use of this device from any place to obtain mobile services which in-turn motivates the security level of the wireless network. The lack of enough security measures leads to the billing disputation among mobile customers, network, and service facilitators. The mobile user must be authenticated

© Springer Nature Singapore Pte Ltd. 2018
Q. Chen et al. (Eds.): ATIS 2018, CCIS 950, pp. 40–51, 2018.
https://doi.org/10.1007/978-981-13-2907-4_4

whenever the user visits the unvisited network via the mobile user's home network. AKA is used foremost in mutual authentication between communicating parties, which authenticates mobile user/client to a network and network/service provider to the mobile user. The architectures in mobile technologies 3G and beyond incorporate mainly three entities; the Mobile Station (MS), Visitor Location Register/Serving GPRS Support Node (VLR/SGSN), and Home Location Register/Authentication Center (HLR/AuC) [8]. Three nodes are participating in encryption or decryption of data and the authentication process.

In spite of complexity in asymmetric cryptography, it is being deployed to provide security services such as authentication and secure communication. Since the Mobile Station Application Execution Environment (MExE) supports asymmetric cryptography, it favors the operators to use PKI in their systems.

1.1 Public Key Infrastructure (PKI) and Wireless Public Key Infrastructure (WPKI)

PKI is the set of hardware, software, people, policies, and procedures needed to create, manage, store, distribute, and revoke digital certificates based on asymmetric cryptography [15]. In the PKI, if one of the end devices is mobile, then it is called as WPKI. A WPKI consists of the following key elements [4]:

i. Certificate authority (CA), a root of trust to issue digital certificates.
ii. A Registration Authority (RA), which is nominated by CA to issue digital certificates and reduction of CA workload.
iii. A database to store certificate requests and revoked certificates.
iv. A certificate store where issued certificates and the private keys are stored in the local computer.
v. The client (Mobile devices).

1.2 Online Certificate Status Protocol (OCSP)

OCSP is used to verify digital certificates. OCSP is an online revocation policy, unlike Certificate Revocation List (CRL) which is an offline revocation policy [11]. OCSP is an alternative for CRL and is communicated over Hyper Text Transfer Protocol (HTTP) with request/respond to messages. Since, OCSP needs less networks bandwidth, enables certificate status checks in real-time and much less overhead on the client as well as on network.

1.3 N^{th} Degree Truncated Polynomial Ring Units (NTRU) for Encryption and Decryption

Since widely used public key cryptography like RSA or Diffie-Hellman cryptosystems tends to be attacked easily by quantum computers, Lattice-based cryptography emerged. It gained more importance recently as a replacement for current public key cryptosystems. NTRU public key cryptosystem is an open source, lattice-based cryptosystem which is fastest and secure encryption algorithm

[9,12]. NTRU algorithm uses the ring of polynomials, $R = Z[X]/(X^N - 1)$. The polynomials in R have integer coefficients: $a(x) = a_0 + a_1X + a_2X^2 + a_3X^3 + \ldots + a_{N-1}X^{N-1}$ [14], then the product $C(X) = a(X) * b(X)$ is the convolution product of two vectors (a and b) having C a size of N positions.

Notations Used in NTRU:

N: N Dimension of the polynomial ring used (i.e. polynomials are up to degree N − 1).

q: "Big" modulus, usually an integer.

p: "Small" modulus, an integer or a polynomial.

r: In Lattice Based Public key encryption, the encryption blinding polynomial (generated from the hash of the padded message 'M' in Short Vector Encryption Scheme (SVES).

mod q: Used to reduce the coefficients of a polynomial into some interval of length q.

mod p: Used to reduce a polynomial to an element of the polynomial ring mod p.

d_f: An integer specifying the number of ones in the polynomials that comprise the private key value f (also specified as df1, df2, and df3, or as dF).

d_g: An integer specifying the number of ones in the polynomials that comprise the temporary polynomial g (often specified as dG). e: Encrypted message representative, a polynomial, computed by an encryption primitive.

F: In SVES (Short Vector Encryption Scheme), a polynomial that is used to calculate the value f when $f = 1 + pF$.

f: Private key in SVES.

g: In SVES, a temporary polynomial used in the key generation process.

h: Public key.

i: An integer.

m: The message in octet string format, which is encrypted in SVES.

1.4   Key Generation for Signing the Document

1. Input: Integers $N, q, d_f, d_g, B \geq 0$ and the string t = standard or transpose.
2. Generate B private lattice bases and one public lattice bases: $set i = B$, While $i > 0$:
    (a) Randomly choose $f, g \in R$ to be binary with d_f, d_g ones, respectively.
    (b) Find small $F, G \in R$ such that $f * G - F * g = q$
    (c) If $t = $ "standard", set $f_i = f$ and $f_i{}' = F$. If $t = $ "transpose", set $f_i = f$ and $f_i{}' = g$. Set $h_i = f_i{}^{-1} * f_i{}' \, mod q$. Set $i = i - 1$.
3. Public output: The input parameters and $h = h_0 \equiv f_0{}^{-}1 * f_0{}' \, mod q$
4. Private output: The public output and the set $f_i, f_i{}', h_i$ for $i = 0 \ldots B$.

1.5   Signing the Document

1. Input: A digital document $D \in D$ and the private key $\{f_i, f_i{}', h_i\}$ for $i = 0 \ldots B$.
2. Set $r = 0$.
3. Set $s = 0$, set $i = B$. Encode r as a bit string. Set $m_0 = H(D||r)$, where $||$ denotes concatenation. Set $m = m_0$.
4. Perturb the point using the private lattices: while $\geqslant 1$:
   (a) set $x = \lfloor -(\frac{1}{q}) * m * f_i{}' \rceil$, $y = \lfloor (\frac{1}{q}) * m * f_i \rceil$, $s_i = x * f_i + y * f_i{}'$
   (b) set $m = s_i * (h_i - h_{i-1}) \bmod q$.
   (c) set $s = s + s_i$, set $i = i - 1$.
5. Sign the perturbed point using the public lattice:
   set $x = \lfloor -(\frac{1}{q}) * m * f_0{}' \rceil$, $y = \lfloor (\frac{1}{q}) * m * f_0 \rceil$, $s_0 = x * f_0 + y * f_0{}'$, $s = s + s_0$.
6. Check the signature:
   (a) set $b = ||(s, s * h - m_0 (\bmod q))||$.
   (b) If $b \geqslant N$, set $r = r + 1$ and go to step 3.
7. Output: The triplet (D, r, s).

1.6   Signature Verification

Same hash function H, norm function $||.||$ and "norm bound" $N \in R$ are required.

1. Input: A signed document (D, r, s) and the public key h.
2. Encode r as abit string. Set $m = H(D||r)$.
3. Set $b = ||(s, s * h - m (\bmod q))||$.
4. Output: Valid if $b < N$, invalid otherwise.

1.7   Encryption

$e \equiv r * h + m (\bmod q)$;
where m: plain text $\in \{-1, 0, 1\}^N$,
$r : random \in \{-1, 0, 1\}^N$,
$h \equiv f^{-1} * g (\bmod q)$.

1.8   Decryption

$a \equiv f * e (\bmod q)$
Lift a to Z^N with coefficients $|a_i| \leqslant \frac{1}{2} q$, $(\bmod p)$ is equal to m.

2   Literature Survey

Zemao et al. [16] explained the optimized authentication and key agreement protocol in the 3G scheme. The key point of the paper was the introduction of the certificate validity ticket generated by the home network which was then sent to the mobile equipment during the authentication process. The authentication between the visited network and home network was facilitated by this protocol.

The security measures such as not accepting the illegal mobile equipment or unauthorized visited network were also taken care of by this.

Bhandari et al. [3] proposed a method that presented the process of Third Generation Partnership Project (3GPP) Authentication and Key Agreement (AKA) protocol and analyzed the mutual authentication which turned to be an essential feature in UMTS standard. This paper explained the advantages of Global System for Mobile communication (GSM) standard for 2G, i.e., using the information as digital signals between Mobile equipment and the service provider. 2G improved the system capacity, but threats were involved in 2G such as weak ciphering algorithms, possible eavesdropping scenario by the false base station. It also explained the need for Third generation (3G) cellular system.

Gururaj et al. [5] described the enhancement of Authentication and Key Agreement by deploying two random keys (Rand1 and Rand2), which was generated by Mobile Station and sent to HLR via VLR. These random keys were used in the process of authentication which turned up the security level; meanwhile, there was a delay in computation speed. The proposed protocol in this paper also reduced the bandwidth consumption between the mobile station and its home network increasing the end to end delay.

El Moustaine et al. [10] implemented NTRU in the authentication for low-cost RFID and explained about the multiplication of polynomials in the ring that in turn results in the convolution product, which was one of the necessary operations done by NTRU. By adapting the NTRU for low-cost RFID tag, it achieved fast encryption and decryption process and high-security level.

Albasheer et al. [1] explained the enhancement of PKI certification validation by making use of NTRU. PKI certification validation and revocation problems were difficult to manage even though the digital certificate was trusted to be the best form of authentication. The chiefly used methods to handle certificate validation and revocation issues were Certificate Revocation List (CRL) and Online Certificate Status Protocol (OCSP). Since RSA (Rivest, Shamir, and Adelman) required high computational power, the NTRU public key cryptographic algorithm was used to replace RSA regarding generation of keys and the digital signature to increase the convenience of mutual authentication service with mobile banking.

Bai et al. [2] explained the analysis of NTRU and its speed of execution with Graphic Processing Unit (GPU). The NTRU key generation, encryption, and decryption were described along with the lattice-based problems such as Shortest Vector Problem (SVP) and Closest Vector Problem (CVP). It was observed that the NTRU performance would increase as the number of devices involved increased.

Park et al. [13] implemented NTRU to protect the payment information in near field communication (NFC) mobile services. It anticipated that the upcoming mobile phones would be provided with NFC interface and, as two NFC devices were brought close to each other, the NFC tag and NFC reader would operate to communicate with each other. Since there was an issue of privacy violation, Zero-knowledge proof based on NTRU was implemented to protect the confidential information.

Micciancio et al. [9] explained the lattice-based cryptography. It differed in the construction of lattice regarding security. The two categories were (a) the one which included the efficient, practical proposals but with less or no proof of security, (b) practical recommendations with provable security but only a few of them could be deployed efficiently in practice. One of the primary and robust features of lattice-based cryptography was the construction of lattice that provided the security guarantee based on the worst-case hardness. The worst-case hardness was nothing but cracking the cryptographic construction. It was as hard as breaking many numbers of lattice problems in the worst case. The main advantage of lattice-based cryptography was the absence of recognized quantum algorithms for solving lattice problems.

Munoz et al. [11] evaluated and compared the certificate revocation policies namely: Online Certificate Status Protocol, which came under online revocation policies and the Certificate Revocation List, which came under offline revocation policies. It explained the need for Online Certificate Status Protocol over Certificate Revocation List since OCSP required low bandwidth than CRL requests. The trade-off between bandwidth and risk level was observed in CRL. Zhao et al. [17] proposed a cache-based OCSP request to get the status of the certificate in real-time with low cost and higher security. The proposed scheme was able to avoid the fake attack on mobile equipment and the server side. It also prevented the denial-of-service attack (DoS) on server side.

Jiang et al. [7] proposed a novel method for anonymous batch authentication using HMAC. The proposed method overcame the issues faced by normal CRL validation such as storage space and checking time, by using Hash Message Authentication Code (HMAC). HMAC was best used for integrity check, as the integrity of the message played a crucial role in the authentication process.

3 Methodology

Figure 1 represents the transmission of messages in the proposed AKA protocol, where ME: Mobile Equipment, VN: Visited Network, HN: Home Network and M1–M5: Messages.

In this research work, the enhanced model of AKA is proposed and implemented. For simplicity, message formats and notations are derived from [16]. The key ideas are:

i. The Foreign Agent and Home Agent authenticate mutually by PKI.
ii. The Home Agent issues the attestation ticket to Foreign Agent only after Foreign Agent gets authenticated successfully.

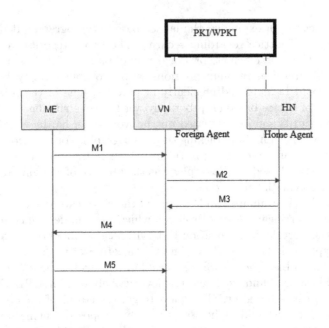

Fig. 1. Transmission of messages in the proposed Enhanced Authentication and Key Agreement (EAKA) protocol

3.1 Notations Used in Enhanced AKA Protocol

URL_X: It denotes address location of digital certificate of entity X;

$DCert_X$: Digital certificate of entity X;

$PU_Y(X)$: Public key encryption of message 'X' using 'Y' as key;

$SK_Y(X)$: Decryption of message 'X' using 'Y' as key;

EMSI: It denotes the Ephemeral Mobile Station Identity number;

N_0: It denotes the Nonce (combination of IP, MAC address of the device and Timestamp);

N_1: It is the encrypted form of N_0;

K_{MH}: It denotes the symmetric secret between ME and HN;

N_V: It denotes nonce generated by VN (combination of IP, MAC address of the device and Timestamp);

Token: It denotes the validity ticket to indicate whether VN is valid;

K_{MV}: It denotes the symmetric key between VN and ME;

$H(M)$: It denotes message digest for message M;

A||B: Message Concatenation;

FA: Foreign Agent;

HA: Home Agent;

FAID: Foreign Agent Identity;

HAID: Home Agent identity;

$SE(X)$: Symmetric encryption of entity X.

3.2   Description of the Enhanced Protocol

The message exchange in the proposed protocol is represented as follows.

- (M1) ME $\rightarrow$ FA: EMSI, FAID, $SE(URL_{ME})$, N_1
  Where: $N_1 = \{N_0\}_{K_{MH}}$;
  N_0 is a nonce generated by ME.
- (M2) FA $\rightarrow$ HA: $DCert_{VN}$, HAID, URL_{ME}, N_1, N_V, H(M)
  Where: M $= PU_{VN}(\text{HAID} ||URL_{ME}||N_1||N_V)$;
  N_V is a nonce that FA generates.
- (M3) HA $\rightarrow$ FA: $DCert_{HN}$, FAID, Token, K', H(M)
  Where: $K' = PU_{VN}(N_V||N_0||K_{MV})$;
  $Token = \{FAID||lifetime||K_{MV}||H(N_0)\}_{K_{MH}}$;
  lifetime is the validity time of Token;
  M $= DCert_{HN}||FAID||Token$.
- (M4) FA $\rightarrow$ ME: TMSI, $Token$, $\{N_2||N_0\}_{K_{MV}}$, $H(M)$
  Where: M $= \text{TMSI}||Token||\{N_2||N_0\}_{K_{MV}}$;
  N_2 is one more nonce that FA generates.
- (M5) ME $\rightarrow$ FA: $\{ack\}_{K_{MV}}$.

3.3   Working of the Enhanced Protocol

Step 1 (M1): ME generates a nonce N_0, which is then encrypted by the key K_{MH} to obtain N1. Where, K_{MH} is a secret key that is pre-shared with its HA. The symmetric key algorithm used is AES since it is proved to be faster and secure among the family. Thumbprint of ME certificate is used to generate the key K_{MH} using AES algorithm. ME sends the EMSI, FAID, Symmetric key encrypted URL_{ME} and N1 to FA. The encrypted URL_{ME} prevents any potential forging. The hash of N_0 is generated using BLAKE hashing algorithm (efficient hash algorithm in the family) and stored for future use to identify modifications. The hashing of MAC address will prevent the impersonation of the malicious device.

Step 2 (M2): FA receives M1, and it queries the Home Agent of ME. It sends the message M2 to HA, by public key encryption of the message M using the NTRU algorithm (public key of FA is used). The hash of the message, H(M) is generated (using the BLAKE hashing algorithm).

Step 3 (M3): The Message, M2 is handled by HA as follows:

i. Rely on the PKI to verify the $DCert_{VN}$ of the foreign network.
   if $Cert_{VN}$ is valid
   {Verify whether message M2 has been modified during the transmission and take necessary action.}
ii. Use the pre-shared secret key K_{MH} to decrypt N_1 and get N_0. HA can optionally verify $DCert_{ME}$ by URL_{ME} and its validity by normal PKI authentication.

iii. Generate the session key for ME and FA for future communication. Denote it as K_{MV}.
iv. Generate the validity token for FA and denoted by Token.
v. Encrypt $N_V \| N_0 \| K_{MV}$ with PK_{VN}, denote the result as K'.

A Response to FA is sent by a message M3. M3 includes message M comprising of $DCert_{HN}$, the identifier FAID, Token and K'.

Step 4 (M4): The message M3 is handled by FA as follows.

i. Construct the message M.
ii. Generate H'(M) and compare it with the received H(M) to check the integrity of the message. If the message is unaltered, test the validity of $DCert_{HN}$ relying on normal PKI.
iii. Decrypt K' with SK_{VN} to get N_v, N_0, and K_{MV}.
iv. Produce a new nonce N_2 and encrypt $N_2 \| H(N_0)$ with K_{MV}. The encrypted value s is represented as $\{N_2 \| H(N_0)\}_{KMV}$. The message M4 is sent to ME, which includes EMSI, Token, $\{N_2 \| H(N_0)\}_{KMV}$ and, hash of the message.

Step 5 (M5): Once the message M4 is received by ME, it retrieves the identifier FAID, $H(N_0)$, and K_{MV} by decrypting the Token by KMH. The ME verifies whether M4 has been changed during transmission using K_{MV}. If the message is intact during transmission, use K_{MV} to decrypt N_2 from $\{N_2 \| N_0\}_{KMV}$, and then send message M5 to VN.

Step 6: After receiving message M5, FA decrypts it to check if the result is ack. If the test succeeds, an ME is admitted to its network. Otherwise, send failure notification and take appropriate action. The protocol fails the authentication and terminates the whole EAKA process if any of the following occurs:

i. The validation of $DCert_{HN}$, $DCert_{VN}$, or $DCert_{ME}$ is unsuccessful.
ii. The Modification detection code is being altered during transmission.
iii. The received Token is not valid or has lifetime elapsed.
iv. Consistency check of the nonce is failed.

4   Testing of the Proposed Method

The Test environment consists of smart phone and laptops. The specification of the mobile client is shown in Table 1. The Certificate Authority (CA) and server details are given in Table 2.

Table 1. Test environment (Client specification)

Device	Samsung galaxy ON5 Pro (Mobile Phone)
OS	Android 6, Marshmallow
RAM	2 GB
Processor	Cortex, 1.3 GHz, quadcore
Software	Android studio 2.3

Table 2. Test environment (CA and Server specification)

Device	HP Probook 440 G4 (Laptop)
OS	Windows 8.1 Enterprise and Ubuntu 16.04
RAM	4 GB
Processor	Intel R(Core), TMi5-52000 cpu @2.2 GHz
Software	OpenSSL, JDK, Python

5   Results

5.1   Performance Comparison of Cryptographic Algorithms in the Proposed EAKA Method

The time taken to execute the proposed EAKA with the NTRU algorithm is less than the time taken to achieve the same with the RSA and ECC (Elliptic Curve Cryptography) algorithm as Fig. 2. Since NTRU is known to be fastest and secured asymmetric algorithm [6,12], the proposed EAKA protocol proves to be more efficient.

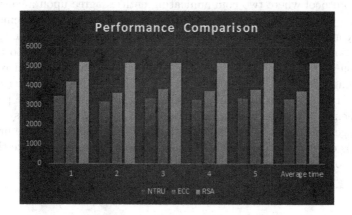

Fig. 2. Execution time comparison

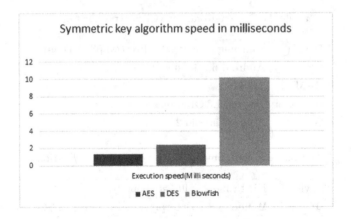

Fig. 3. Symmetric key algorithm performance

Figure 3 shows the performance of symmetric key algorithms in the proposed EAKA scheme. The AES algorithm has minimum execution time required and is found to be faster. Hence, it is selected for symmetric key encryption of messages in the EAKA scheme.

6   Conclusion

A novel and efficient approach for authenticated key agreement for mobile networks called EAKA is proposed. Entity Authentication is a process by which an entity gains confidence in the communicating entity. It is often coupled with the distribution of "session key" with communicating entity. The session key generated is used later to achieve the confidentiality and integrity. Key agreement is a protocol where two communicating entities agree upon a shared key. If it is done precisely, it prevents an undesired third party from forcing a key choice on communicating parties. The proposed scheme increases the speed of authentication process more securely. The proposed work takes in to account the creation of the validity token at the home agent which helps in reducing the signalling overhead. A comparison is made between contemporary asymmetric key cryptographic algorithms in terms of execution time in the proposed EAKA scheme. NTRU is chosen as it is proved to be faster. The symmetric key exchange is achieved using the AES algorithm. The efficacy of the proposed algorithm is compared with OPAKA scheme. The results demonstrate that the proposed protocol is more secure and fast.

References

1. Albasheer, M.O., Bashier, E.B.M.: Enhanced model for PKI certificate validation in the mobile banking. In: 2013 International Conference on Computing, Electrical and Electronic Engineering (ICCEEE), pp. 470–476, August 2013
2. Bai, T., Davis, S., Li, J., Jiang, H.: Analysis and acceleration of NTRU lattice-based cryptographic system. In: 15th IEEE/ACIS International Conference on Software Engineering, Artificial Intelligence, Networking and Parallel/Distributed Computing (SNPD), pp. 1–6, June 2014
3. Bhandari, R., Choudhary, A.: 3GPP AKA protocol: simplified authentication process. Int. J. Adv. Res. Comput. Sci. Softw. Eng. 4(12), 655–658 (2014)
4. Forouzan, B.A.: Cryptography and Network Security, 1st edn. McGraw-Hill Inc., New York (2008)
5. Gururaj, H.L., Sunitha, R., Ramesh, B.: Enhanced authentication technique for next generation 3GPP networks. In: 2014 International Conference on Contemporary Computing and Informatics (IC3I), pp. 1361–1365, November 2014
6. Hoffstein, J., Pipher, J., Schanck, J.M., Silverman, J.H., Whyte, W., Zhang, Z.: Choosing parameters for NTRUEncrypt. In: Handschuh, H. (ed.) CT-RSA 2017. LNCS, vol. 10159, pp. 3–18. Springer, Cham (2017). https://doi.org/10.1007/978-3-319-52153-4_1
7. Jiang, S., Zhu, X., Wang, L.: An efficient anonymous batch authentication scheme based on HMAC for VANETs. IEEE Trans. Intell. Transp. Syst. 17(8), 2193–2204 (2016)
8. Khan, W., Ullah, H.: Authentication and secure communication in GSM, GPRS, and UMTS using asymmetric cryptography. Int. J. Comput. Sci. Issues (IJCSI) 7(3), 10 (2010)
9. Micciancio, D., Regev, O.: Lattice-based cryptography. In: Bernstein, D.J., Buchmann, J., Dahmen, E. (eds.) Post-Quantum Cryptography, pp. 147–191. Springer, Berlin (2009). https://doi.org/10.1007/978-3-540-88702-7_5
10. El Moustaine, E., Laurent, M.: A lattice based authentication for low-cost RFID. In: 2012 IEEE International Conference on RFID-Technologies and Applications (RFID-TA), pp. 68–73, November 2012
11. Munoz, J.L., Forne, J., Castro, J.C.: Evaluation of certificate revocation policies: OCSP vs. Overissued-CRL. In: Proceedings of the 13th International Workshop on Database and Expert Systems Applications, pp. 511–515, September 2002
12. Nguyen, H.B.: An overview of the NTRU cryptographic system. Ph.D. thesis, San Diego State University (2014)
13. Park, S.W., Lee, I.Y.: Anonymous authentication scheme based on NTRU for the protection of payment information in NFC mobile environment. J. Inf. Process. Syst. 9(3), 461–476 (2013)
14. Shen, X., Du, Z., Chen, R.: Research on NTRU algorithm for mobile Java security. In: 2009 International Conference on Scalable Computing and Communications; Eighth International Conference on Embedded Computing, pp. 366–369, September 2009
15. Stallings, W.: Cryptography and Network Security: Principles and Practice, 6th edn. Prentice Hall Press, Upper Saddle River (2013)
16. Zemao, C., Junge, Z., Biyi, H.: Optimizing PKI for 3GPP authentication and key agreement. In: 2012 Fourth International Conference on Multimedia Information Networking and Security (MINES), pp. 79–82. IEEE (2012)
17. Zhao, X., Wenyan, Z., Shanshan, C.: New certificate status verification scheme based on OCSP for wireless environment. In: 2009 International Forum on Computer Science-Technology and Applications, vol. 2, pp. 195–198, December 2009

Information Abuse Prevention

Information Abuse in Twitter and Online Social Networks: A Survey

M. J. Akshay[1]([⊠])(iD), N. Ala Mohammed[2], Amit Vyas[2], Palayil Baby Shony[2], and Sheldon Kevin Correa[2]

[1] Department of Information and Communication Techonolgy,
MIT, Manipal, MAHE, Manipal, India
akshay.mj@manipal.edu
[2] Department of Computer Science and Engineering,
MIT, Manipal, MAHE, Manipal, India
avmn666@gmail.com, 23amitvyas@gmail.com, shony.baby@yahoo.com,
sheldoncorrea97@live.com

Abstract. With the Internet accessible to more than half of the global population, online social network, especially Twitter, has become a highly impactful source of news and information. However not all the information spread through Twitter are credible. Spammers, scammer, people with agendas and so on may use the services of Twitter, making it difficult for the users to tell the difference between what is credible and what is not. Current researchers work on several ways to monitor the tweets and the users who spread falsehoods. This paper aims to study a few papers in this field and that employs the usage of several algorithms as well as human intuition to track credibility of tweets and also the different ways the features of Twitter and other online social network are exploited to spread misinformation by certain users for their own benefits and profits.

Keywords: Credibility · Misinformation · Online social network
Twitter

1 Introduction

As of the first quarter of 2018, Twitter has nearly 336 million active users and about 500 million tweets generated per day. These huge numbers indicate why Twitter is such a huge and impactful medium of information in this modern age. However, its popularity and reach makes it a breeding ground for misinformation and spams. Therefore, it is necessary to study how tweets can be categorized as credible, how they can be exploited for various wrongful purposes and also study how they propagate across the internet.

Online Social Networks (OSNs) allow people to voice their opinions as well as create, edit and view content. This definitely has made us technologically superior but at the same time has also given rise to the issue of widespread misinformation. The growth of online social networks made it possible for a significant

Q. Chen et al. (Eds.): ATIS 2018, CCIS 950, pp. 55–66, 2018.
https://doi.org/10.1007/978-981-13-2907-4_5

revolution in the domain of information dissemination and communications. The popularity of OSN have led to make the source of information more reliable and trustworthy. Because of the rumors spreading in OSNs or misinformation, trust relations mostly gets exploited and could make ways that could potentially cause bad effects. In order to maintain the trustworthiness or dependability of online sources, it is vital to limit the viral effect of misinformation.

Twitter grants a verification symbol to pages/people who are considered credible. But a great majority of the users are normal people whose credibility or aims can't be verified. Twitter can be an effective way to spread wrong information about someone or some organization and there are considerable chances that some or many of the audience would believe such messages (tweets). This could be particularly effective in case of political campaigns where Twitter could be used to glorify a candidate which the user supports and also smear an opposition candidate, both by using contents that are not facts. Tweets supported by images that are out of context or edited could prove even more effective although there isn't any real truth behind them.

A lot of misinformation is spread across the internet, the identified types include incomplete information, pranks, contradictions, out of date information, improperly translated data, software incompatibilities, unauthorized revisions, factual errors, biased information and scholarly misconduct. Popular social networking platforms like Facebook and Twitter have recently become a major news source and also has become a platform for viral marketing. With the increased benefits comes the drawbacks like spread of misinformation which could potentially lead to undesirable effects and because of the huge online social networks (OSN) there comes a threat like a user spreading the viral misinformation on such networks.

Another way to misuse the features of Twitter is spamming. Spammers could take advantage of political campaigns or popular events based on Twitter's Trending Topics to spread links which gullible users would click. This may either have monetary benefits in the form of on-click-ads and shopping sites or have content that helps propagate certain agendas.

The easy access of Twitter via web and mobile application, the simple format of tweets (only 280 characters; previously 140), ease of propagation (in the form of Retweets) and so on aids these malignant entities. Recently, attackers and spammers have also started to exploit ways in which search results on various search engines can be manipulated so as to reach on top of related search result pages.

2   Literature Survey

Twitter is a social media that has attracted great interests from computer scientists and researchers. Twitter maintains several APIs that allow the collection of data regarding the tweets to be much easier. Several studies have used these features to make in-depth analysis on the credibility of tweets, how falsehood spreads through the medium and the average user's response to such fake information.

2.1 Credibility Ranking of Tweets During High Impact Events

Gupta et al. in [1] makes an important observation on how major world events and the tweets based on them could be used by certain users to spread misinformation and sometimes spam links as well. It was the first work in this topic that used automatic ranking techniques to evaluate credibility at the most basic level of information. Logistic linear regression analysis was used to obtain good predictors for credibility of the tweets. The performance of Rank-SVM and PRF using the NDCG evaluation metric was used to evaluate the ranking.

Data Collection: Twitter's Streaming API was used for data collection. This API allows us to extract real-time tweets, using query parameters such as the words in the tweet, time of posting and so on. Trends API from Twitter was used to obtain query terms, that returns top 10 trending topics on Twitter. Events that had at least 25,000 tweets were taken. The additional criteria was that the event should have been trending for at least 24 h.

Advantages: (i) The combined set of features of context and source based analysis have produced significant statistical improvement. (ii) With the help of regression analysis and ranking algorithms, we can categorize the credible tweets and users. Automated algorithms that employs superior machine learning and relevance feedback approach based on Twitter features makes it more efficient to assess credibility of tweets.

Disadvantages: (i) Use of manual annotation process is time-consuming and also may produce subjectively varying outputs.

2.2 Measuring User Influence in Twitter: The Million Follower Fallacy

Cha et al. in [2] studies the concept of influence of directed links in Twitter. These directed links determine the flow of information and therefore they are an indicator of a user's influence on others. How influence diffuses through people is quite unknown and questions such as if it varies across topics and over time is unanswered. This paper aims to answer such questions, especially in the field of marketing. Marketing entities actively search for potential influencers to promote products. The advertised items are usually well outside of the domain of expertise of these hired individuals. We need to know the effectiveness of these marketing strategies. We also need to question if a person's influence in one area is applicable to other areas.

Data Collection: The activities or features through which a user can influence others on twitter are mainly categorized under three measures- Indegree (number of followers), Retweets and Mentions. Activities of 6 million users were monitored. For each of these users, the value of each influence measure was computed and compared. Instead of comparing the values directly, relative order of users' ranks was taken as a measure of difference. Under each measure of influence, the users were ranked. Spearman's rank correlation coefficient was used to measure of the degree of association between two rank-sets. The top 20 from each category was taken. Also, the relative influence ranks for the three measures were also tabulated.

Advantages: (i) The exact behaviours that help a normal user to gain vast amount of followers could be characterized.
(ii) The influence pattern in Twitter was presented as an empirical analysis.
(iii) The results gives some major insight into viral marketing and social media design.

Disadvantages: (i) "Influence" has been defined. In a variety of ways. It is not clear what exactly influence means. This could make the final results very subjective.
(ii) There is a lack of data that could be used to test the hypotheses or the 'influentials' theory from previous studies.

2.3   From Obscurity to Prominence in Minutes: Political Speech and Real-Time Search

Metaxas et al. in [3] makes an important study on how the real-time search results feature provided by popular search engines could be exploited by malignant tweeters to spread falsehoods on a larger scale and that too instantaneously. Appearing on the first search page gives such false news an opportunity to spread virally, which otherwise they wouldn't have had. This can be exploited by political campaigns to spread their agendas, especially during the time of elections. New information on blogs and Twitter can quickly feature at the top of search results which helps their motives.

Data Collection: For this study, a 2010 US Senate election contested by Scott Brown and Martha Coakley was used. Around 50,000 tweets were collected for the study (albeit some portion of this had to be ignored due to time constraints).

Advantages: (i) Analyzing the signature of spam attacks will prove to be important in the future as it will help to build mechanisms that can automatically detect these types of attacks.
(ii) The reply-activity of users was visualized using the force-directed algorithm. This efficiently lets us visualize thereby helping in making hypotheses and analyses.

Disadvantages: (i) Most of the assumptions made in the study tries to quantify the intentions of human interaction. Therefore there could be reduction in accuracy of the results.

2.4 From Information Credibility on Twitter

Castillo et al. in [4] try to focus on various automated methods to assess the credibility of a set of tweets in this paper. More trending posts are analyzed and their credibility is determined based on extracted features. Here, supervised learning is used and a dataset is built based on studying the credibility of information on Twitter by human assessors. Twitter monitor was used to detect over 2500 cases and an event supervised classifier was run over these cases where 4 levels of credibility were considered. Then Automatic credibility analysis is performed based on 4 types of features: message-based features, propagation-based features, user-based features and topic-based features.

Data Collection: For In this study, a collection of discussion topics relevant to the study were extracted by studying bursts of activity. Then, the labelling of the topics was done by a batch of human evaluators according to whether it belongs to "news" or "casual conversation'. After the creation of the dataset, the credibility of the items in the former class was assessed by another group of judges. Twitter Monitor (which detects sharp rises in the frequency of sets of keywords was used to detect Twitter events over a 2 month period. A labeling round was done to separate the topics which spread the information and facts about a news event, from those which mainly relate to personal opinions or chat.

Advantages: (i) The results show that there exist considerable differences in the ways that messages tend to propagate, which can be used to automatically classify them as credible or not, with precision of 70%–80%.

2.5 Tweeting Is Believing? Understanding Microblog Credibility Perceptions

Morris et al. in [5] presents a survey in this paper regarding users' perceptions of credibility of tweets. An investigation done on the perceptions of credibility of tweets is presented and the results obtained in a survey about various features that tend to have an impact in the assessments of tweet credibility by users are reported. A collection of 26 features was used to design the survey. It was established that Participants tend to come upon tweets through search which draws out more of a concern about their credibility as compared to encountering tweets of the people or organizations followed. Online experiments were conducted where numerous properties were changed to understand the influence on the perception of the reader. The features were narrowed down to a set of 3 based on the influence they had on the perceptions of the readers and also based their visibility to the readers on Twitter and the search engines. User Name, Message Topic, and User Image were the features selected. Analysis was done to test the influence of the experiments on the perceptions. It was determined that the users are more worried about the credibility of the tweets when they are not obtained from their followed users which results in them making judgements in situations such as search, based on the presented information.

Data Collection: An initial study was conducted where users were instructed to "think aloud" while conducting a search on Twitter. They were observed and questions were asked during this process. The features mentioned by the participants were noted and a collection of 26 features obtained were used in designing the survey. The participants were asked about the specific features and they were also asked to rate the influence of the features on credibility.

Advantages: (i) This survey shows that the tweets encountered through search results have a higher influence on people rather than encountering the tweets by those followed. So, this study shows what the users perceive about the credibility of tweets which could be beneficial in future research related to misinformation on twitter.

Disadvantages: (i) There might be errors in determining the credibility since some of the interface features might be hidden due to the fact that they are unnecessary while consuming content.

2.6   Twitter Under Crisis: Can We Trust What We RT?

Mendoza et al. in [6] proposes that in order to keep a track on whether the information is correct or not it becomes necessary to analyze the propagation of the tweets. The aim of this paper was to conduct analysis on the activity associated to the earthquake that took place at Chile in the year 2010 and describe the twitter activity in the hours and days following this incident. On plotting the data on graph, a logarithmic format was obtained in both axes. The classification of tweets is done and is split into several categories i.e.: affirms, denies, questions. The categories also include the unknown and the unrelated. The classification results demonstrates that a large percentage i.e. approximately 95.5% of tweets related to confirmed truths validate the information (i.e. the category which contains "affirms") whereas the number or percent of tweets that rejects these valid cases is significantly low i.e. around 0.3%.

Data Collection: In the hours and the days that followed the incident of earthquake, by the use of Twitter enabled to tweet very important time-critical information such as alerts, safe locations and the services available. The tweets were associated to 716,344 distinct users who have an average of around 1018 followers and 227 followees.

Advantages: (i) The credibility of tweets is critical in times of a disaster and this paper provides an insight into that.

2.7   What Is Twitter: A Social Network or a News Media?

The primary objective of this paper is to show the topological characteristic of the online social media - twitter and its influence which serves a medium for dissemination of information. The goal is to obtain the characteristics through user profiles, social relations and tweets. At first, the social network is analyzed and various studies have been conducted regarding certain key factors such as the distribution of followers and following, the association between the followers and tweets, degrees of separation, reciprocity and homophily. Users are then ranked by the number of tweets posted by them, their followers, number of retweets and pageRank. Then comes the analysis of trending topics where these topics are classified, information on the number of users that account for their participation in the topic is obtained. Finally the information sharing is studied through the analysis of retweets where retweet trees are constructed and their temporal and spatial characteristics are examined.

Data Collection: According to this paper, the user profile information (full name, location, biography, users followed by etc.) is collected. The method followed to collect user profiles - the team started with the consideration of a popular figure who has over one million followers and then considered the information of his/her 'followers' as well as of the 'following'. Around 41 million of user profile data was collected in this similar fashion. Then the top topics currently trending are collected via the twitter search API. The API returns the title of the topic that is trending, a query in the form of a string, and the time of the request. The string obtained from the API was used to grab all the tweets that mention the particular topic. 262 distinctive trending topics and their associated tweets were collected. 1500 tweets for each query were returned which is the maximum number. Thereafter these tweets are filtered for any possibility of containing spam in them.

Advantages: (i) This research helps to understand the influence what information can have with regard to the user that handles it, whether or not the topic is trending, how retweeting can become a powerful aspect etc. These data also gives us information in order to study the human behavior.

3   Literature Survey Model Review

With a large amount of tweets generated every other second, it is important to classify which among them the users can trust and which all they can ignore.

3.1   Credibility of Information

This is the core discussion that this study aims to provide. [1] and [4] provides us their final results regarding the validity of tweets. [1] takes into account only high-impact events around the globe. 14% of the tweets were found to contain spam links. 30% were regarding situational awareness, of which only 17% was credible. [4] considered only tweets that were classified as "news". Table 1 shows the final results obtained by [4].

These percentages show a different picture to that of the results from [1]. In [1], a very small portion (17%) was credible situational information while a comparable amount (14% was spam. However, in [4] a much higher amount of tweets (41%) have been classified as credible. From this, we could draw the conclusion that viral or high-impact events can be breeding grounds for fake information and spams when compared to tweets that provide general information or news

Table 1. Credibility of tweets related to "news"

Category	Percentage
Almost certainly true	41.0
Likely to be false	31.8
Almost certainly false	8.6
Ambiguous	18.6

3.2   Validation

A unique angle to the question of Twitter's credibility was offered by [6] where they try to validate confirmed truths using Twitter during the period of a disaster. Their results show that 95.5% of the time, facts are validated whereas only 0.3% they are denied. An interesting inference from this is that the Twitter community behaves like a collaborative information filter.

3.3   Influence, Homophily and Reciprocity

Homophily is a tendency for similar people to have more contact among them than with dissimilar people. Reciprocity is when a follower of a user gets followed back by that user.

Papers [2] and [7] looks into the extent of influence, homophily and reciprocity among Twitter users. [2] brings into spotlight the factor of "influence". It uses three measures - indegree, retweets and mentions - to determine how influence can be used to promote marketing strategies. With a higher indegree (i.e. higher number of followers), a user can influence more people with his/her ideas or tweets. Their spatial analysis shows that most influential users can have influence to a great extent across a variety of topics. The most popular Twitter accounts had a disproportionate amount of influence, which was followed a power-law distribution. How various types of influentials interact with their audience was shown by the temporal analysis done in [2]. [7] gives us an in-depth look into how homophily and reciprocity exists in Twitter. Its results show that Twitter has major diversions from the most common traits of social networks: distribution of followers does not obey power-law, the degree of separation has a shorter diameter than expected, and most of the relationships are not reciprocated. However, if we consider reciprocated relationships, amongst them there exists a certain level of homophily.

3.4   Real-Time Search Results and User Perception

Most modern browsers provide what is known as "real-time search results". Contents from microblogs, news sites, Twitter, etc., could get featured on top of the search results within moments of their generation. This means that Twitter acts not just as a social medium but also as source for news. [3] and [5] both agree with this and makes their study based on the effects of real-time results [3] considers a US Senate election and how false agendas were spread during the days leading up to the voting day. Spammers with political motives exploited this feature of the search engines. This study was able to find multiple spam accounts that were created for this very purpose. These tweets could reach normal users who aren't followers of the tweeter or not even using twitter. The study shown in the survey paper [5] takes into account this effect of tweets - how they can serve as popular sources of news rather than just a social media with users having limited reach (within their followers). The results from [5] indicate that users who arrive at a tweet from a search result are likely to be more skeptical than users who are the followers of the tweeter. Such users are thus expected to make credibility judgments depending on the information they possess.

4 Literature Survey on Additional Papers Related to Online Social Networks

4.1 Determining Difference Between Knowledge and Belief

Grier et al. in [9] determines the difference between knowledge and belief is other significant problem because belief is easily mistaken for knowledge. People may become so convinced of their own expertise in a particular area that contradictory incoming information is dismissed without consideration. This problem links critical thinking to epistemology, and illustrates the uncertain demarcation between prior knowledge, beliefs, and bias. Components of evaluation process:

Metacognition: Metacognition is knowledge or cognition that takes as its objects or regulates any aspect of any cognitive endeavor. Metacognitive regulation is vital because the thinker should actively think for the information and think again whether it is correct or not.

Disposition: It is attitude of being disposed to consider in a thoughtful, perceptive manner the problems and subjects that come within the range of one's experiences. There are three ideas in favor of disposition: that some people by nature are more likely to evaluate; that people criticize most ideas as a matter of course; and conversely, that people must be selective about the ideas they choose to criticize.

4.2 Reverse Diffusion

Budak et al. in [10] explains the role of reverse diffusion. Reverse diffusion method traces back to find the origin of the misinformation. This could get increasingly difficult when more number of users, networks and nodes are in a complicated, tangled mesh. In ranking-based, Imeter algorithm can be used. The attack originates from somewhere and later spreads to other nodes. If so, we can definitely apply the reverse process to trace back the origin of attack. It is this intuition on which this algorithm is based upon. The Reverse Diffusion Process (RDP) is employed. In this, active (affected) nodes are taken and the reverse flow is studied at each node/step, eventually leading to the origin node. A key observation from Reverse Diffusion Process is more the node is hit by a reverse flow more likely it is an attacker. The number of hits for a node is gauged as an Imeter value. Higher the Imeter value, higher chances of a node being the attacker.

4.3 Social Diffusion Model

Domingos et al. in [12] discusses the social diffusion model. It is used to analyze information, misinformation, and disinformation. Milieux - Development of information, misinformation, and disinformation is around a person's cultural, social and historical milieu/milieux. In this model the information, misinformation and disinformation are looked upon based on the influential factors like social, cultural and historical contexts. This leads to the diffusion of the information.

Diffusion: Information, misinformation and disinformation diffuse through the internet or across social groups. The content is shared even if people might not believe it or it may not be recognized as deceptive or inaccurate. Diffusion may be rapid or slow depending on the situation of the content. Unknowns: The next element in this model focuses on the unknowns. The pure intent behind diffusion is an unknown factor. The diffusion of misinformation or disinformation could be driven either for a hostile or a generous cause.

Deception: In the creation of disinformation, the objective of the deceiver is to deceive. Deception does not guarantee success in achieving objectives, whether personal or social motivation. It could be antagonistic or benevolent or sometimes neither of them. Due to the existence of variety of objectives show why deception is highly complex.

Judgement: Judgements of the believability is then made by the receiver using cues to credibility or cues to deception. These cues to credibility can also be used even by the deceivers in order to deceive the receivers. Much of the judgment depends on the level by which the receivers suspect information based on their knowledge. This part of the model represents the association of the information literacy behavior throughout this diffusion model process.

5   Conclusion

By studying these papers, we have understood the different ways in which the features, the popularity and the reach of Twitter could be misused for various purposes and we have also seen the various methods used to detect the misinformation present. These can be in the form of spams, scams or just false information. These studies reveal that a considerable portion of tweets about news, people or events are not credible or verified information.

   In this paper online social network have been discussed about the methods of analyzing the propagation of misinformation. The most important thing to know is the difference between the knowledge and belief. If a user have a knowledge of something than it would be difficult to misinform that person whereas if someone believes that the source is right than it is easy to hack the source and flood the internet with incorrect or misinformation.

The reverse diffusion helps in tracking back to the source from where the misinformation started. With this process one can find the source leading to misinformation and stop following such source. Immunization is computationally expensive but helps to find the leading cause which lead to the propagation of misinformation on the internet. Finally the social diffusion models helps to analyze misinformation, information and disinformation using which a person can finally judge the source and find out whether the source is reliable or not. Since the social network is a fast growing network the information propagates very fast whether it is correct information or incorrect information and to avoid having incorrect information it is necessary that the correct information spreads through the online social network so that by the time incorrect information reaches a user the user is already aware of correct information.

6   Future Scope

With the ever-increasing popularity of Twitter, more studies are necessary to figure out the credibility of tweets. Twitter provides an open API thereby making new researches to collect data more easily. These researches can help the developers of Twitter to make further updates to curb the spread of spam and malicious links and also educate the users when or how to ignore tweets that are likely to be fake. Findings that the Twitter community acts as a truth-filter, it gives rise to a new research field which tries to find possibilities to detect rumors using methods such as aggregate analysis of tweets.

References

1. Gupta, A., Kumaraguru, P.: Credibility ranking of tweets during high impact events. In: Proceedings of the 1st Workshop on Privacy and Security in Online Social Media, Lyon, France. ACM (2012)
2. Cha, M., Haddadi, H., Benevenuto, F., Gummadi, K.P.: Measuring user influence in twitter: the million follower fallacy. In: Proceedings of the 4th International Conference on Weblogs and Social Media (ICWSM), Washington DC, USA. AAAI (2010)
3. Mustafaraj, E., Metaxas, P.T.: From obscurity to prominence in minutes: political speech and real-time search. In: WebSci10: Extending the Frontiers of Society On-Line, Raleigh, USA. The Web Science Trust (2010)
4. Castillo, C., Mendoza, M., Poblete, B.: Information credibility on Twitter. In: Proceedings of the 20th International Conference on World Wide Web, Hyderabad, USA. ACM (2011)
5. Morris, M.R., Counts, S., Roseway, A., Hoff, A., Schwarz, J.: Tweeting is believing?: Understanding microblog credibility perceptions. In Proceedings of the ACM 2012 Conference on Computer Supported Cooperative Work, Raleigh, USA. ACM (2012)

6. Mendoza, M., Poblete, B., Castillo, C.: Twitter under crisis: can we trust what we RT? In: Proceedings of the First Workshop on Social Media Analytics, Washington DC, USA. ACM (2010)
7. Kwak, H., Lee, C., Park, H., Moon, S.: What is Twitter, a social network or a news media? In: Proceedings of the 19th International Conference on World Wide Web. ACM, Raleigh, USA (2010)
8. Mathioudakis, M., Koudas, N.: TwitterMonitor: trend detection over the twitter stream. In: Proceedings of the 2010 International Conference on Management of Data, pp. 1155–1158. ACM (2010)
9. Grier, C., Thomas, K., Paxson, V., Zhang, M.: @spam: the underground on 140 characters or less. In: Proceedings of the 17th ACM Conference on Computer and Communications Security, 04–08 October 2010, Chicago, Illinois, USA (2010)
10. Budak, C., Agrawal, D., El Abbadi, A.: Limiting the spread of misinformation in social networks. In: Proceedings of the 20th International Conference on World Wide Web, WWW 2011, New York, NY, USA, pp. pp. 665–674. ACM (2011)
11. Goyal, A., Bonchi, F., Lakshmanan, L., Venkatasubramanian, S.: On minimizing budget and time in influence propagation over social networks. Soc. Netw. Anal. Min. **3**, 1–14 (2012)
12. Domingos, P., Richardson, M.: Mining the network value of customers. In: Proceedings of the Seventh ACM SIGKDD International Conference on Knowledge Discovery and Data Mining, pp. 57–66, 26–29 August 2001, San Francisco, California (2001)

Location Privacy Protection for Sink Node in WSN Based on K Anonymous False Packets Injection

Ling Song[1,2(✉)], Wei Ma[1,2], and Jin Ye[1,2]

[1] Guangxi University, Nanning 530004, China
jqian@gxu.edu.cn
[2] Guangxi Key Laboratory of Multimedia Communications and Network
Technology, Nanning 530004, China

Abstract. In wireless sensor networks, the convergence node (sink node) is the center of the network, all network data will be transmitted to the sink node, it will process and extract effective information, which leads to the uneven distribution of traffic. An external attacker who can monitor traffic will find the location of the convergent node and attack it according to this feature. In order to protect the location of the sink node in the wireless sensor network, a privacy protection protocol based on K anonymous false packet injection (KAFP) is proposed. The protocol randomly generates K false sink nodes, transmits real data to the sink node and transmits false data to the false sink node. By hiding the location of the sink node, the security time is increased. Theoretical analysis and simulation experiment results show that KAFP can protect privacy of convergent nodes at lower energy consumption when the value of K is properly selected.

Keywords: Wireless sensor network · Sink node · Location privacy protection
K anonymous · False packet injection

1 Introduction

Wireless sensor network (WSN) [1] is an important part of the Internet of things, a large number of micro sensor nodes are formed by self-organization. At present, it is widely used in traffic management, disaster warning, medical and health, national defense, military, environmental monitoring, industrial manufacturing, and many other fields, help people get more accurate information.

However, in the process of practical application, wireless multi-hop communication is used to transmit messages in WSN, which is vulnerable to attack by attackers, and thus causing serious security problems. Therefore, privacy protection in WSN has become an important research direction. The existing privacy protection of WSN can be divided into two categories: Data privacy protection [2] and location privacy protection [3]. Data privacy protection technology mainly adopts privacy protection technologies such as perturbation, anonymity and encryption [4], it achieves the task of data aggregation, data query and access control without revealing the privacy information; Location privacy protection technology can be divided into base station location privacy protection and source location privacy protection. For attackers to acquire source location or base station location information by monitoring communication mode, probabilistic flood routing, phantom routing, fake packet injection, camouflage real source node or base station, loop trap routing [5] and other protection strategies are adopted, they can effectively prevent network sensitive location information from leaking, at the same time, control the consumption of network energy and reduce communication delay.

This paper focuses on location privacy protection of the sink node, a privacy protection protocol based on K anonymous false packet injection (KAFP) is proposed.

2 Related Work

Sink node attackers can also be divided into two types:global traffic attackers and hop by hop attackers. The global traffic attackers have large storage space, strong computing power and sufficient energy to monitor the whole network. According to the characteristics of large traffic flow near the sink node, the attackers can quickly locate around the sink node and find the convergent node by searching one by one. The hop-by-hop tracking attackers track to the convergent node from the node where the message is forwarded, along the direction of the flow of information.

Deng et al. [6] proposed that when the traffic in the network is small or there is no traffic, the false attackers are sent to deceive the attackers and send them to the direction far away from the sink nodes. In the document [7–10], the author proposes that the false packet is injected into the network at the same time, and the transmission of the false packet is in the direction of the far neighbor node to reduce the traffic in the network hot zone, so it can better protect the location privacy of the sink node and increase the security time of the network. In the scheme of document [7], the amount of data forwarded and received for all nodes is consistent, so that the global traffic attackers can attack the network nodes through traffic analysis. The scheme also puts forward the fusion of false packet transmission and multipath routing. The traditional privacy protection protocol packets are only transmitted on a specific path, and multipath transmission, as the name implies, has multiple transmission paths. The three privacy protection schemes proposed by Chen [11], such as bidirectional tree, dynamic bidirectional tree and zigzag two-way tree, are all based on multi-path. Directional walking can also be used to resist attackers' attacks on nodes. Document [12] proposes a RBR strategy, which consists of three parts: first building ring routing and ring-shaped interregional routing. Secondly, the data in the network is sent to the nearest ring route through the shortest path. Finally, the nodes on the ring route are going along the ring route direction and the vertical ring route direction respectively. Data packets are forwarded to get better privacy protection capability. Nezhad and other [13] proposed a method of hiding the location of the sink node. This method uses all nodes to broadcast route discovery information to hide the location of the sink node, but it also brings about the problem of excessive network energy and the danger of congestion.

3 System Model

3.1 Network Model

- The network uses a planar network structure to deploy sensor nodes, and only one sink node is responsible for communicating with the outside world and exchanging information.
- The nodes in the network periodically send their collected information to the sink nodes, and update the real-time status of the monitoring targets at any time.
- Both the source node ID and the destination node ID are included in the data field of the packet in the network. The source node ID of all nodes refers to the data source of the node, that is, the ID of the last hop; because the final purpose of the node is to transfer the data to the sink node, so the destination node ID refers to the location of the sink node.
- The confidentiality of data, the application of cryptographic algorithms and the secure processing of communication channels are not discussed in this paper.

3.2 Attacker Model

Attackers are attracted by huge commercial interests, locate and capture the sink nodes, and have strong attack ability. In this paper, we assume that the attack model has the following properties:

- The main task of the attacker is to get the location of the sink node so as to get the whole network information stored in the sink node. Therefore, it does not need to attack the network nodes actively or modify the transmission packets, and does not need to interfere with the normal operation of the network, but instead of passive tracking and eavesdropping, and stealing information related to the sink nodes.
- Attackers have large storage space, strong computing power and sufficient energy to monitor the entire network.
- By the order of data transmission, the attacker is tracked by hop by hop, starting from the source node with information forwarding, following the flow direction of the information, tracking the hop and hop to the convergence node, at this time the location of the source node is assumed to be safe.
- The monitoring radius of the attacker is the same as the communication radius of the sensor nodes in the network.

4 KAFP Protocol

4.1 Basic Ideas

In wireless sensor networks, sink nodes serve as the hub of the network, connecting external networks and sensor networks. The external network collects the network information to obtain useful data by integrating the sink nodes. The sensor network transmits the information collected by the source nodes in the network to the sink nodes through multiple hops. Therefore, as a key node, the sink node has a connecting role. If the sink node is attacked by an attacker, the whole network will not run normally and face the risk of data loss and network reconstruction. In order to protect the location of the sink nodes effectively, the K anonymous sink nodes are used to simulate the traffic flow of the real sink nodes, increase the search scope of the attackers and improve the network security time.

4.2 Routing Strategy

Network Initialization. Firstly, the sink node sends out the broadcast message. Each node will increase one hop every time the message arrives, so that all the nodes in the network update their hop to the sink node in time and produce their own near neighbor nodes. Because the location of the sink node in the network is known, K false sink nodes are randomly generated around the aggregation nodes. When the sink node is broadcast messages, all nodes get their own jump number to the sink node, and the false sink node also performs the same operation, so that all nodes in the network can also get their own to false sink nodes and the number of hops and the nearest neighbor nodes to the false sink node.

Transmission of Data Packets. After network initialization, the source node sends packets to the sink node and selects the next hop from the nearest neighbor node. This ensures that the real data is closer to the sink node and transfers data to the sink node in the shortest path.

Transmission of False Packets. When the network is initialized, K anonymous nodes are randomly generated. When the source node sends real data to the sink node, the false packet is sent to the false sink node. The format and size of the false packet are the same as the real data packet. When the transmission path of the false packet coincides with the real packet transmission path, the transmission of the false packet is abandoned, and the transmission path is reclosing as long as there is a transmission node reclosing. Because the object of the false packet is ID as a false sink node, the destination ID of the real packet is a sink node. By the purpose of the packet's ID, it can be used to judge whether the packet is a real packet or a false packet. When the real packet has been forwarded, then the packet is given up when the packet is received.

As shown in Fig. 1, the node is approximately uniformly distributed in the network, and the source node source selects the next hop from its near neighbor node to transmit the data to the sink node in the shortest path, and then the false packet is also transmitted from the source node to the false sink node. When the false packet is transmitted to the false sink2 node, when the first node is transmitted to the post source node, if the true packet sent to the sink node has passed, the false packet gives up the transmission; if the true packet to the sink node is not passed, the false packet continues to transmit, and can continue to transmit when the back true packet arrives. The whole process only happens after the real packet is transmitted, and the path coincides before discarding.

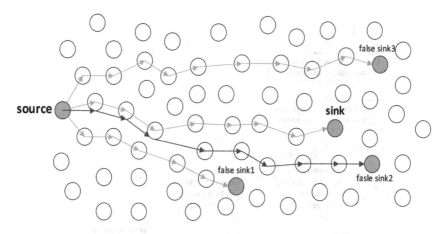

Fig. 1. Routing strategy

Sink nodes can receive real data, false sink1 and false sink3 can receive fake packets. The red transmission path in the graph shows only the path that the false packet should go when the shortest path is transmitted. If the true packet is transmitted, the false packet is discarded when the first node after the source node is transmitted.

4.3 Working Process

The work flow of the KAFP protocol is shown in Fig. 2:

(1) all nodes in the network get location information of sink nodes and false sink nodes.

(2) through the message broadcast node, you can get your nearest neighbor node.

(3) the source node sends data packets to the sink node and the false sink node. If the data is sent to the sink node, the next hop is selected from the nearest neighbor node in accordance with the shortest path algorithm, until the packet arrives at the sink node and the task ends; if the data is sent to the false sink node, the (4) step is executed.

(4) when the information is sent to a false sink node, if the node has a real packet through it, the data is given up and the task is finished; if the node does not pass the real packet, the (5) step is executed.

(5) when the node does not pass the real data packet, the next hop is selected from the nearest neighbor node in accordance with the shortest path algorithm, until the false packet arrives at the false sink node and the task is over.

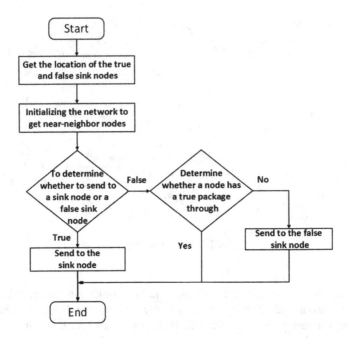

Fig. 2. Work diagram of KAPF

5 Performance Analysis

5.1 Analysis of Communication Overhead

The communication overhead of KAFP is mainly divided into three parts, namely, communication overhead generated by network initialization, communication overhead generated by real packet transmission and communication overhead generated by false packet transmission. The communication overhead generated by network initialization includes the broadcast cost of sink nodes and the broadcast overhead of false sink nodes. The cost of sink nodes is related to the literature. It only needs to consider the broadcast cost of false sink nodes. The communication cost of real packet transmission is the number of nodes that the data traversed from the shortest path to the sink node from the source node.

The communication cost of real packet transmission is the number of nodes that the data traversed from the shortest path to the sink node from the source node. The communication overhead of the false packet is divided into two cases: when the false data passes from the source node to the false converge node, the node of the false packet is completely different from the real packet, and the communication overhead of the false packet is the number of hops passing through the node. When the false data passed from the source node to the false converge node is the same as that of the real packet, the false packet is discarded and the transmission stops, thus reducing the use of the network traffic in the same area, and reducing the communication overhead accordingly. The communication cost of the whole network is the sum of the communication cost of the real packet and the communication cost of the false packet.

5.2   Analysis of Safety Performance

In the process of network initialization, K false sink nodes are randomly generated, and the positions of false sink nodes are deployed by normal distribution. The basic idea is as follows: first, the system selects a minimum value d_{min}, which represents the shortest distance between the true sink and the false sink node. Using d_{rand} to represent the true distance between sink nodes and false sink nodes, it is obvious that $d_{rand} \geq d_{min}$. We choose the false sink node in the circle with the sink node as the center and the radius d_{rand}. We use the following formula to calculate d_{rand}:

$$d_{rand} = d_{min} \times (|x| + 1) \tag{1}$$

Among them, X is randomly generated and obeys the normal normal distribution, that is $X \sim N(0, 1)$. So, the density function $\varphi(x)$ is:

$$\varphi(x) = \frac{e^{-\frac{x^2}{2}}}{\sqrt{2\pi}} (x \in R) \tag{2}$$

The distribution function $\Phi(x)$ can be obtained by the density function:

$$\Phi(x) = \int_{-\infty}^{x} \frac{e^{-\frac{t^2}{2}}dt}{\sqrt{2\pi}} \tag{3}$$

Therefore, the probability of d_{rand} in the interval $[d_{min}, \rho d_{min}]$, where, P is:

$$P = P(-\rho \leq x \leq \rho) = \Phi(\rho) - \Phi(-\rho) = \Phi(\rho) - (1 - \Phi(\rho)) = 2\Phi(\rho) - 1 \tag{4}$$

The reason that the probability of d_{rand} in the interval $[d_{min}, \rho d_{min}]$ is 4 is: suppose $\rho = 1$, if there is no offset d_{rand}, then the probability of d_{rand} in the interval $[-1, 1]$ is $2\Phi(1) - 1 = 0.6826$. Considering that d_{rand} refers to the actual distance, the probability of the interval $[-1, 1]$ is equivalent to the probability in a circle with $(0, 0)$ a center and a radius of 1, which is expressed in the interval $[0, 1]$. When there is an offset d_{min}, it is also assumed that $\rho = 1$, then the probability of d_{rand} in $[0, 2d_{min}]$ is $2\Phi(1) - 1 = 0.6826$, and d_{rand} refers to the actual distance and $d_{rand} \geq d_{min}$, so the probability in the interval $[0, 2d_{min}]$ is equivalent to the probability of a circle with $(0, 0)$ a center with a radius of $[d_{min}, 2d_{min}]$, which is expressed in the interval $[d_{min}, 2d_{min}]$.

Therefore, the probability of d_{rand} falling in the interval $[d_{min}, 2d_{min}]$ is $2\Phi(1) - 1 = 0.6826$; the probability of falling in interval $[d_{min}, 3d_{min}]$ is $2\Phi(2) - 1 = 0.9544$; and the probability of falling in interval $[d_{min}, 4d_{min}]$ is $2\Phi(3) - 1 = 0.9974$. As shown in Fig. 3, A, B, C, D, E, F, G represent randomly generated 7 false sink nodes, the probability that they fall between the interval $[d_{min}, 4d_{min}]$ is 0.9974, and the nodes that can be seen as the normal distribution will fall within the $[d_{\min}, 4d_{\min}]$ range. However, the nodes H and I in the range of d_{min} can not be false convergence nodes.

Therefore, we can choose the communication radius of the d_{min} for the convergence node, so that the false converging node will not affect the transmission of the data of the converged node outside the communication radius of the converged node. Attacker tracking the safe time of the sink node is related to the number of anonymous nodes K, K is several, and security is increased several times.

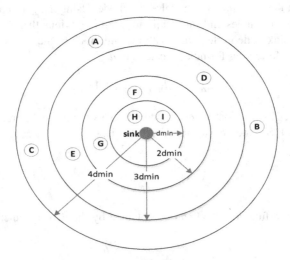

Fig. 3. Selection of false sink node

6 Simulation Experiment and Analysis

6.1 Comparison of Communication Overhead

In this paper, the simulation experiments are carried out under the MATLAB platform. 900 sensor nodes are randomly deployed in the monitoring area of 400 m * 400 m. The sink node is located at the center with a coordinate of (200, 200), the source node is located at (400, 400), and the communication radius of each node is 20 m. 100 times of simulation were carried out, and the average value of 100 experiments was taken as the result.

In this paper, the average number of packets transmitted from the source node to the base station is used as the indicator of the communication overhead. Due to the close relationship between the number of nodes and the energy consumption, we have carried out experiments on the communication overhead under the condition that the number of false sink nodes is different. Because the anonymous nodes are generated by normal distribution, the locations of the pseudo sink nodes are centered in the concentric nodes with a radius greater than 20 m and less than 80 m. From Fig. 4, we can see that with the increase of anonymous nodes, the communication cost will also increase.

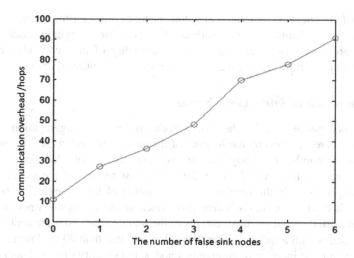

Fig. 4. False sink number and communication overhead

Now take K = 3, that is, when there are 3 anonymous nodes, the KAFP, the location privacy protection routing protocol (LRP) [14] and the secure zone based node location privacy protection routing protocol [15] (SD-DRP) are compared in terms of communication overhead in the same scene. Figure 5 shows the corresponding communication overhead of the distance of the distance converging nodes of the source node to 5, 10, 15, 20, 25, 30, 35 respectively. With the increase of hop count between the source node and the sink node, the communication cost of the three protocols increases. Compared with the LPR protocol and the SD-DRP protocol, the KAFP communication overhead is reduced by 4% and 16.5% respectively. In the LPR protocol, each network node has a probability of generating false packets. In the SD-DRP

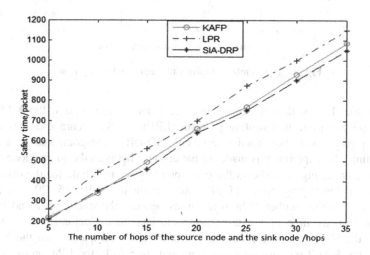

Fig. 5. Comparison of communication overhead

protocol, a false packet is generated in a non secure region to form a ring area, which all leads to a large communication overhead. However, the anonymous nodes selected by KAFP are fewer, and there may be the possibility of path coincidence, so the communication overhead is less than the two protocols mentioned above.

6.2 Comparison of Safety Performance

When the first packet sent by the source node arrives at the aggregation node, the security time is represented by the K times of the number of packets sent by the source node (K is the number of anonymous nodes). The number of false sink points also affects the security time, which is similar to the last section of the communication overhead experiment. In the case of different number of false converging nodes, the experiment is carried out on the safety time. Because the anonymous nodes are generated by normal distribution, the locations of the pseudo sink nodes are centered in the concentric nodes with a radius greater than 20 m and less than 80 m. From Fig. 6, we can see that with the increase of anonymous nodes, the security time will also increase.

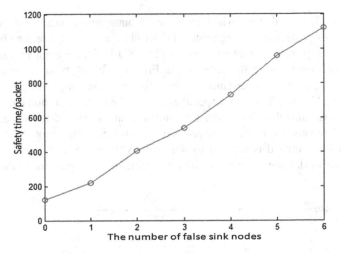

Fig. 6. The number of false sink nodes and safety time

Now take K = 3, that is, when there are 3 anonymous nodes, the KAFP, the location privacy protection routing protocol (LRP) and the secure zone based node location privacy protection routing protocol (SD-DRP) are compared in terms of security time. An experiment is made for the distance between the source node and the converging node. Figure 7 shows the corresponding security time for the distance of the distance converging nodes of the source node to 5, 10, 15, 20, 25, 30, 35 respectively. As the number of hops increases between the source node and the sink node, the security time of the three protocols increases. Compared with the LPR protocol, the KAFP security time is reduced by 12%, compared with the SD-DRP protocol, the KAFP security time has increased by 5%. In the LPR protocol, every network node has probability to produce false packets, which makes it difficult for attackers to attack and locate the sink nodes. Therefore, the security time of KAFP is lower than that of the LPR protocol.

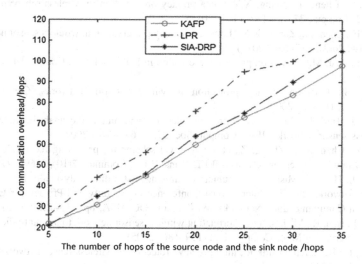

Fig. 7. Hops between the source and the sink and safety time

7 Summary

In this paper, a protocol of sink node privacy protection based on K anonymous false packet injection is proposed. The protocol hides the real sink nodes through multiple anonymous nodes and increases the communication path to imitate the transmission process from the source node to the sink node by the communication path. The area with low network traffic is fully utilized. At the same time, the number of anonymous nodes is also an important factor affecting the privacy protection effect. The more anonymous nodes, the better the protection effect on the sink nodes, but the communication cost will also increase. The anonymous nodes should not be too much, otherwise the network will inevitably cause congestion because of excessive energy consumption. Theoretical analysis and simulation experiments show that, under a limited number of anonymous nodes, the protocol proposed in this paper increases a portion of the communication overhead, but it is still considerable for the improved security time. Therefore, this protocol has better privacy protection performance.

Acknowledgments. This work is supported by the National Natural Science Foundation of China (nos. 61762030).

References

1. Ren, F.Y., Huang, H.N., Lin, C.: Wireless sensor networks. J. Softw. **14**(7), 1282–1291 (2003). (in Chinese with English abstract)
2. Fan, Y.J., Chen, H., Zhang, X.Y.: Data privacy preservation in wireless sensor networks. Chin. J. Comput. **35**(6), 1132–1147 (2012)
3. Peng, H., Chen, H., Zhang, X.Y.: Location privacy preservation in wireless sensor networks. J. Softw. **26**(3), 617–639 (2015)
4. Qian, P., Wu, M.: Survey on privacy preservation in IoT. Appl. Res. Comput. **30**(1), 14–21 (2013)
5. Wang, S.: Location privacy protection in wireless sensor networks. Central South University, HuNan (2009)
6. Deng, J., Han, R., Mishra, S.: Intrusion tolerance and anti traffic analysis strategies for wireless sensor networks. IEEE Comput. Soc. **32**(5), 637–646 (2004)
7. Jian, Y., Chen, S.G., Zhang, Z., et al.: A novel scheme for protecting receiver's location privacy in wireless sensor networks. IEEE Trans. Wirel. Commun. **7**(10), 3769–3779 (2008)
8. Deng, J., Han, R., Mishra, S.: Countermeasures against traffic analysis attacks in wireless sensor networks. In: First International Conference on Security and Privacy for Emerging Areas in Communications Networks, Washington, DC, USA, pp. 113–126 (2005)
9. Deng, J., Han, R., Mishra, S.: Decorrelating wireless sensor network to inhibit traffic analysis attacks. Pervasive Mob. Comput. **2**(2), 159–186 (2005)
10. Zhang, K.: Research on sink-location privacy protection in wireless sensor networks. Anhun University (2014)
11. Chen H.L., Lou, W.: From nowhere to somewhere: protecting end-to-end location privacy in wireless sensor networks. In: Performance Computing and Communications Conference, Albuquerque, New Mexico, USA, pp. 1–8 (2010)
12. Li, Z.: An energy-efficient and preserving sink-location privacy scheme for WSNs. Central South University, HuNan (2012)
13. Nezhad, A.A., Makrakis, D., Miri, A.: Anonymous topology discovery for multi-hop wireless sensor networks. In: Proceedings of the 3rd ACM Workshop on Qos and Security for Wireless and Mobile Networks, New York, USA, pp. 78–85 (2007)
14. Yang, Y., Shao, M., Zhu, S., et al.: Towards statistically strong source anonymity for sensor networks. ACM Trans. Sens. Netw. **9**(3), 34 (2013)
15. Li, P.Y., Zhang, Z.X.: Protocol algorithm design of location privacy preserving for Sink node based on security area. J. Syst. Simul. **27**(12), 2973–2980 (2015)

Sentiment Analysis to Enhance Detection of Latent Astroturfing Groups in Online Social Networks

Noora Alallaq[1(✉)], Muhmmad Al-khiza'ay[1(✉)], Mohammed Iqbal Dohan[2], and Xin Han[3]

[1] School of Information Technology, Deakin University, 221 Burwood Highway, Burwood, VIC 3125, Australia
{nalallaq,malkhiza}@deakin.edu.au
[2] Department of Multimedia, College of Computer Science and Information Technology, University of Al-Qadisiyah, Diwaniyah, Iraq
mohammed.iqbal@qu.edu.iq
[3] College of Computer Science, Xi'an Shiyou University, Xi'an, China
xhan.xsyu@outlook.com

Abstract. Astroturfing is a sponsored and organized campaign for making a favor to the sponsor by the act of expressing an opinion via reviews on a product or service. Such reviews are misleading and cause potential risk to the public in general as there has been increasing trend of making decisions based on the online reviews. In this paper, we proposed an enhanced LDA to have a novel model known as Latent Group Detective Model (LGDM) which is meant for solving the problem of astroturfing group detection. This model is based on the fact that there is a difference between a generative process of normal review and a potential astroturfing review. LDGM accommodate a label for finding sentiment associated with reviews to have the better analysis of AGs. Four cases are considered for AG analysis. Case 1 observes the groups as they are discovered. Case 2 performs opposite sentiment analysis. Case 3 focuses on high sentiment reviews while case 4 deals with change of sentiment of potential AGs across products or services.

Keywords: Online reviews · Astroturfing · Astroturfing group
LGDM · Sentiment analysis

1 Introduction

The invent of Web 2.0 has changed the way Internet based applications acquired capabilities to facilitate greater collaboration among enterprises, content providers and end users. This has resulted in social media such as Facebook, Twitter etc [15]. Similarly online application like (http://Yelp.com) that support online reviews and micro reviews provided by customers of different products and

© Springer Nature Singapore Pte Ltd. 2018
Q. Chen et al. (Eds.): ATIS 2018, CCIS 950, pp. 79–91, 2018.
https://doi.org/10.1007/978-981-13-2907-4_7

services. Amazon and Trip Advisor are other examples where product reviews and hotel reviews are made respectively [16]. User-generated online reviews provide rich information related to various products and services, probably, without bias.

This is the reason why such data over World Wide Web (WWW) became goldmine to researchers and enterprises in the real world. The rationale behind this is that mining online reviews can yield business intelligence (BI) to make strategic decisions. Online reviews are able to influence customer purchasing decisions, they became crucial for any business to garner knowledge from them and make well informed decisions. So far so good as there is utility of online reviews for decision making [10]. However, there is another side of the coin that online reviews may be positive or negative and given by malicious people with bad intentions. By giving negative reviews, a product or service is affected as the reviews lead decisions that lead to monetary loss and damage of reputation. Often astroturfers are paid for doing so. The group involved in astroturfing is known as Astroturfing Group (AG). Generally this group is latent but involved in the reviews that are fake with organized effort. Detecting such group can have high utility in combating astroturfing [7].

Many researchers contributed towards detection of astroturfing reviews with different models such as author model, topic model and topic-author model. Often they used a generative statistical model in Natural Language Processing (NLP) known as Latent Dirichlet Allocation (LDA) as explored in [6,13]. Of late, we extended topic-author model to have group topic-author model for efficient identification of AGs. our work assumed certain factors related to latent AGs. The assumptions include (a) astroturfing is made like a paid campaign; (b) it has temporal dimension as it lasts only for few days; (c) astroturfing reviews are made in a given context; (d) astroturfing is made on a similar topic; (e) sentiment might exist in the tweets. Our work is similar to this work [15] but it lacks thorough sentiment analysis including patterns.

To overcome this, this paper focuses on identification of astroturfing reviews, identifying potential AGs and analyzes them in terms of evaluating that given AG is really the group behind astroturfing campaign. AG with itself is compared among different hotel reviews to ascertain the change in sentiment. In this paper we proposed a generative process model known as Latent Group Detective Model (LGDM) which is meant for solving the problem of astroturfing group detection. This model is based on the fact that there is difference between generative process of normal review and a potential astroturfing review. The LGDM is employed not only to discover AGs but also the sentiment exhibited by them. Towards this end AGs are analyzed with four cases. The first case finds AGs based on similarity in topic, context and time posted. In addition to this, the second case considers sentiment associated with reviews as well. In addition to case 2 parameters, case 3 considers high polarity in sentiment associated with AGs. The fourth case exhibits the AGs across the hotels with dominating sentiment such as positive or negative. The following are our contributions.

- A generative process model named LGDM is proposed and implemented for finding AGs from the datasets collected from (http://Yelp.com). This model considers sentiment aspect of the underlying reviews in addition to similarity in topic, context and time posted.
- Sentiment analysis is made to analyze AGs well. We considered four cases for this purpose. In the first case, AGs are analyzed with similarity of topic, context and time of posting without considering sentiment. It results in finding number of AGs across the topics specified. The topics considered are food, price, location and service. In the second case sentiment is also considered along with similarity of topic, context and time posted. This leads to know AGs to find opposite sentiment. The third case is to know AGs that do have high sentiment polarity while the fourth case deals with AGs that are associated with hotels and have overall positive or negative campaign.

The remainder of the paper is structured as follows. Section 2 reviews relevant literature on astroturfing and sentiment analysis. Section 3 methodology. Section 4 experiment setup. Section 5 experiment result. Section 6 evaluation. Section 7 concludes the paper besides giving possible directions for future work.

2   Related Works

Fake accounts and fake product reviews or astroturfing became evident in social networks. This section reviews related works on this issue and sentiment analysis for extracting trends or patterns of misbehaviour.

2.1   Astroturfing and Anomaly Detection

Astroturfing and anomaly detection mechanisms are essential to fight against misleading reviews and fake identities over OSN. Towards this end, Savage et al. [11] proposed a methodology for detecting anomalies in online social networks. It has two stages. As fake accounts are used to make fake reviews and influence genuine users, Viswanath et al. [13] on the other hand proposed unsupervised anomaly detection techniques along with Principal Component Analysis (PCA). Then they used ground-truth data provided by Facebook to evaluate their work. Mukherjee et al. [8] studied the problem of opinion spamming. They opined that fake reviewer group is very dangerous as they can influence sentiment of people in large scale.

Hooi and Akoglu [5] studied online ratings fraud and provided useful insights. They stated that fake reviews can be detected by using their characteristics such as appearance in short bursts of time and skewed rating distributions. Aklogu and Faloutsos [1] provided a review of methods that are employed in fraud detection in OSNs.

Crowdturfing is nothing but malicious crowd sourcing which became a security problem that cannot be ignored. Song et al. [12] proposed a framework known as CrowdTarget to achieve target based detection of crowdturfing. It detects target objects associated with crowd sourcing such as URL, page and post. Mukherjee et al. [7] proposed a fake review detection technique based on classification which distinguishes real reviews from pseudo reviews.

2.2   Role of Sentiment Analysis

Sentiment analysis is the phenomenon in which text corpora is subjected to a procedure which can classify the textual data into positive, negative and neutral. However, much can be done using sentiment analysis including identification of trends in sentiment, emotion, and kind of emotion like anger besides severity. [9] proposed a method to analyze sentiment of customers against various brands like KLM, IBM and so on. With sentiment analysis they could identify most famous and preferred brands by general public. They employed subjective and objective polarity evaluation for both positive and negative sentiments. They evaluated their approach with ground truth sentiment analysis results and found it to be effective. [2] explored competitive analysis of social media by pizza industry to have useful knowhow on decision making. [3] made a survey of techniques used for sentiment analysis. They include classification approaches that are based on supervised learning. Machine learning techniques such as SVM, k-Nearest Neighbour, and Rule Based Classifier (RBC) etc. are studied in terms of their merits and demerits.

The problem of astroturfing and detection of user groups behind it can provide latent information about malicious reviews and their authors. However, it is more useful when sentiments reflected in the reviews are mined and trends or patterns associated with astroturfing are comprehended. Mostafa [6] proposed a method to analyze sentiment of customers against various brands like KLM, IBM and so on. He et al. [4] explored competitive analysis of social media by pizza industry to have useful knowhow on decision making.

Zhao et al. [18] proposed a sentiment analysis system that works for Chinese corpus. It was named as MoodLens that maps emotions with different sentiments. Cambria et al. [3] explored multi-model sentiment analysis that considers multimedia including facial expressions in images or videos to understand sentiments. Wang et al. [14] on the other hand used USA president election cycle of 2012 for real time sentiment analysis from Twitter tweets. Their proposed method takes public tweets on USA elections in for analyzing public sentiment towards different political parties.

There is significant necessity to work on astroturfing group detection so as to uncover organized groups participating in astroturfing. Moreover, there is need for discovering sentiment trends associated with their astroturfing activities to have more useful business intelligence for making well informed decisions on fraud incidents.

3 Methodology

In this section, we formally define astroturfing group detection problem. Then we describe the proposed model known as Latent Group Detective Model (LGDM) which is a probability graphical model. The model is based on the fact that there is difference between generative process of normal review and a potential astroturfing review.

3.1 Problem Difinition

Astroturfing takes place when a group of astroturfers are employed by a company to spread certain sentiments, on its products or services, which are not actually real. Astroturfing is therefore an organized effort which spreads opinions intentionally through different channels. Astroturfer groups are hired to support the exaggerated claims of their employer and bring positive opinion of people in favour of employer. At the same time they may work to deny adverse claims. In this context, online review is considered to be a major vehicle for expressing opinions and spread sentiment among people in favour of employer of astroturfing groups.

Let N_p denote the number of products in the dataset. The products are denoted as $P = \{p_1, p_2, p_3, \ldots, p_{N_p}\}$. The number of reviewers spreading sentiment for a single product is denoted as N_u. Thus the reviewer and the corresponding review are represented by x_{ij} and r_{ij}, where $j = 1, 2, 3, \ldots, N_u$. we assume there are t latent astroturfing groups $g = \{g_1, g_2, \ldots, g_t\}$ in the observed dataset. For each group g_k, $k = 1, 2, 3, \ldots, t$, it contains a set of reviewers who are considered suspicious astroturfing. Accordingly, the objective is: from all reviewers x and their corresponding reviews r, to effectively detect the number of astroturfing groups t and the reviewers x form the g_k. Table 1 summarizes the notations used in this section.

3.2 Latent Group Detective Model

Astroturfing leads to positive opinions on a target product for its promotion or negative opinions on the same for demotion. However, with respect to astroturfing, the sentiment associated with each review made by a reviewer is controlled by a group. On the other hand the sentiment associated with a review of a genuine reviewer is said to be intuitive. From this it is very clear that the generation process associated with astroturfing review is different from that of normal review. In each latent astroturfing group, all group members are said to have same behaviour with respect to review generation process.

3.3 Modelling the Generation Process

As mentioned earlier, the LGDM is a generative probabilistic model to represent generation of set of reviews. With respect to latent group assignments, a set of

Table 1. Notations used in Latent Group Detective Model [LGDM]

Notation	Description
N_w	The number of words in a review
N_p	The number of products in the dataset
N_u	The number of reviewer in a single product
G	The number of latent groups
g	The group assignments for all reviewers
θ	The G dimension group distribution
α	The Dirichlet prior for group assignments distributions
x	A reviewer of a product
ϕ	Reviewer Distribution associated with a group
β	The Dirichlet prior for word. Distributions
l	The sentiment label of a review
z	The sentiment distribution associated with a group
π	The Dirichlet prior for sentiment distribution
w	A word in a review
γ	Words distribution associated with a sentiment label and a reviewer
φ	The Dirichlet prior for words distribution

reviewers are represented as random mixtures. Each latent astroturfing group is characterized by its distribution among reviewers. The structure of LGDM contains boxes or plates reflecting replicates while the observed data is represented by dark shaded nodes. Yet other nodes in the graphical model are nothing but latent or unobserved data.

LGDM assumes the following generative process for each review R: The model first picks a group assignment g from a multinomial distribution θ. Then, according to the picked group g and a multinomial distribution ϕ, the reviewer x is generated. Meanwhile, for the picked group, a sentiment label l (positive or negative) for the target product is drawn from a multinomial distribution z. Afterwards, the generated reviewer x generates a review R according to the designed sentiment label l. As for the review R, each word w of it is independently drawn from a distribution γ defined by l and x. Finally, as a Bayesian generative model, we give each multinomial distribution θ, ϕ and γ, a prior distribution in the generative process. When we consider the complete model, the reviewer set is dynamically divided into G groups, and each of which contains a number of reviewers. For each group, the genuine reviewer is allocated this group with a very lower probability. Then a preselect threshold is used to filter the genuine reviewer. Therefore the group only contains the reviewers which are most likely belong to it after filtering.

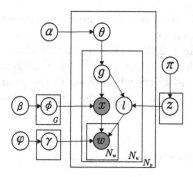

Fig. 1. Graphical model representation of Latent Group Detection Model

Solution Procedure

In the Latent Group Detective Model, for a product p_i, given the hyper parameters α, β, π and φ, and a set of N_u reviewers x_{ij}, the joint distribution of a reviewer mixture ϕ, a sentiment label l, a sentiment label mixture z, the R_{ij} presented by a set of N_w words w is given by:

$$p(R_{ij}, \theta, g, l, \phi, \gamma, z \mid \alpha, \beta, \varphi, \phi, x_{ij}) = p(z \mid \pi)p(\phi \mid \beta)p(\gamma \mid \varphi)$$

$$\times \prod_{j=1}^{N_u} \prod_{k=1}^{N_w} p(w_k \mid \gamma)p(x_{ij} \mid \phi)p(g \mid \theta)p(l \mid z)$$

Integrating over ϕ, γ, θ, z and summing over g and l, we get the marginal distribution of a review of a product.

$$p(R_{ij} \mid \alpha, \beta, \varphi, \phi, x_{i_j}) = \iiiint p(z \mid \pi)p(\phi \mid \beta)p(\gamma \mid \varphi)$$

$$\times \prod_{i=1}^{N_u} \prod_{k=1}^{N_w} \sum_g \sum_l p(w_k \mid \gamma)p(x_{ij} \mid \phi)p(g \mid \theta)p(l \mid z)\mathrm{d}\phi\mathrm{d}\gamma\mathrm{d}\theta\mathrm{d}z$$

Then we take the product of the marginal probabilities of single review, the probability of all the reviews of one product is:

$$p(R_i \mid \alpha, \beta, \varphi, \phi, x_i) = \prod_{d=1}^{N_u} p(R_{ij} \mid \alpha, \beta, \varphi, \phi, x_i)$$

Finally, we get the probability of all review of all product is:

$$p(R \mid \alpha, \beta, \varphi, \phi, x) = \prod_{i=1}^{N_p} p(R_i)$$

At this point we use Gibbs sampling, a standard approximations method for graphical model [15], to calculate

$$p(g_t \mid x, R_i), t = 1, 2, 3, \ldots, G$$

which is the posterior distribution of the group assignments of a reviewer given the reviewer and their reviews.

For each group g_t, t in $1, 2, 3, \ldots, G$, calculating the value of posterior distribution of g_t over all reviewers, then a preselect threshold, eg. $k = 0.7$, can help us chose the most likely reviewers who belong to the group g_t. As result, we get all members of each astroturfing group. Then, the problem proposed in Sect. 3.1 can have a solution.

4   Experimental Setup

This section provides the settings of experiments made with LGDM proposed in Sect. 3.2. It presents details of dataset and the procedure followed to complete experiments is described. The aim of the experiments is to detect suspicious astroturfing groups besides analyzing the sentiment patterns they employ in the process of astroturfing. Since sentiments influence people either positively or negatively, the LGDM model throws light into not only detecting groups but also the dynamics of sentiments.

4.1   Data Collection

One of the famous tourism social network platforms known as (http://Yelp.com) is used to collect document corpus. Each document is nothing but a review containing opinion of customers or reviewers. The corpus we collected has 10 datasets. Each dataset contains reviews related to a hotel associated with tourism industry. After collecting datasets, we manipulated them to add one more attribute that is a label showing whether a review is fake or real. One more label is required in this paper as we focused on the sentiment analysis to enhance the quality of AG detection. With respect to sentiment we produced sentiment label using SentiWordNet API. This label plays crucial role in sentiment analysis for producing AGs with different characteristics.

4.2   Dataset Modifications

To examine our methodology we added 200 fake reviews and made the reviews associated in the same period of time, similar context and same topic, and distributed them among all datasets, hotels, to detect the same AGs in all datasets to prove that they exist in all data sets. Datasets are subjected to changes in order to have meaningful experiments that achieve 4 cases of AG detection that considers sentiment label as well.

Case 1: Finding AGs based on similar topic, context and time posted.
Case 2: Analyzing AGs that contain opposite sentiment.
Case 3: Analyzing AGs with high sentiment reviews.
Case 4: Analyzing change of sentiments of potential AGs.

For the case 1, datasets are used as they are collected without manipulating sentiment. To meet case 2, the datasets are modified. For this, we select 100 reviews of 200 fake reviews that we created and change their sentiment to opposite the most reviews. For instance, if the most reviews are positive, so we will change those 100 reviews to negative. According to that, the sample result of dataset is as in Table 2 Fig. 2. The LGDM model accommodates topics as well. The topics considered are food, price, location and service. Sentiment is distributed as mentioned above to detect astroturfing groups based on sentiment analysis.

To achieve case 3, we will make the remaining 100 fake reviews as high sentiments. For example, if the most hotel reviews are positive or negative, we will try to find among them the potential AGs by making the 100 fake reviews that we created as high positive if the most hotel reviews are positive and vice versa if negative. According to this, we will detect AGs because they are associated in same topic, context, and topic and have the same high positive sentiment. The sample excerpt of dataset is as shown in Table 3 Fig. 2.

Case 4 is related to finding change of sentiments of potential AGs. In this case the dataset is modified to ensure that there is change of sentiments of potential AGs across hotel datasets. An excerpt of sample dataset is shown in Table 4 Fig. 2.

Table 2. Sample case2 dataset for H1 and T1

User Name	Reviews sentiment	Authentic
U1	Positive	Real
U2	Positive	Real
U3	Positive	Real
U4	Positive	Real
U5	Positive	Real
U6	Negative	Fake
U7	Negative	Fake

Table 3. sample of case3 dataset for H1 and T1

User Name	Reviews sentiment	Authentic
U1	Positive	Real
U2	Positive	Real
U3	High positive	Fake
U4	High positive	Fake
U5	Positive	Real
U6	Negative	Real
U7	Negative	Real

Table 4. all datasets with potential AGs

Hotels	AGs	Sentiment Opinions
H1	50	Positives
H2	50	Negatives
H3	50	Positives

Fig. 2. Samples of dataset cases

5   Experimental Results

We built a prototype application to demonstrate the functionality of the proposed LDGM. Pre-processing of all datasets is done. First of all continuous English words separated. Then all the reviews are subjected to stop words removal. Stop words are the words in English language that do not have their effect in the text processing with respect to finding astroturfing. After removing stop words, a simple PorterStemmer algorithm is used to perform stemming which remove derived words. In fact derived words are reduced to their stem word or the root form of the word. After completion of pre-processing, the datasets are subjected to LDGM process as described in the Sect. 3 The four cases described in Sect. 4.2 are used for experiments.

As showing in Fig. 3 there are three results show AGs detected as (case 1: similar features, case 2: high sentiment, and case 3: opposite sentiment) on each topic are presented with the number of reviewers found in each AG. Each group has different number of reviewers found for each topic. From the results it is understood that the number of reviewers in each astroturfing group differ and there is no common number to generalize this. The results revealed that different topics also have different number of reviewers in each group. The last case is to find AGs among all hotels based on changes of sentiment that presented in Fig. 4.

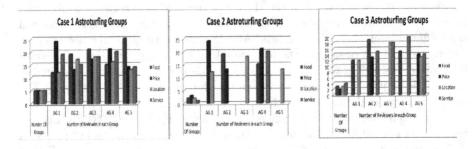

Fig. 3. Results of LDGM cases

Hotel ID	No. of AGs	Sentiment Propagated
H1	20	Negatives
H2	25	Negatives
H3	18	Positives
H4	21	Positives
H5	25	Negatives

Fig. 4. Results of case 4

6 Evaluation

The Latent Group Detective Model (LGDM) proposed in this paper is meant for finding latent AGs. The results are found for four different cases. The AGs are discovered by the application that implemented LGDM. The performance of the model is evaluated as described here. The dataset after pre-processing is randomly partitioned into 5 sub datasets of equal size. Out of the five sub datasets one sub dataset is used as testing set while the remaining sets are used as training sets. The process of cross-validation is repeated for 5 times with each sub dataset one time. F1 score is measured as follows.

$$F1 = 2 * ((precision * recall)/(precision + recall)) \tag{1}$$

Once precision and recall are measures as per the equations, the F1 measure computed. The results are provided as shown below.

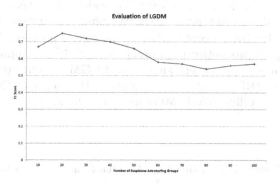

Fig. 5. Evaluation of LGDM

As shown in Fig. 5, the number of suspicious astroturfing groups is presented in horizontal axis while the vertical axis showed the F1 measure. From the 5-fold cross validation, it is understood that F1 Score is the measure based on precision and recall. They are evaluated and presented to know the accuracy of the proposed model. The F1 score visualized in the graph reveal the efficiency of the proposed LGDM.

In the literature the state of the art such as [7] and [17] are close to our work. They focused on finding groups behind fake reviews online. The difference between them is that in [7] fake reviewer group detection is made while the [17] focused on finding clusters of fake accounts in Online Social Network (OSN). They have not used generative statistics model with graphical view. In our work, we defined LGDM based on the generative process model with graphical view as presented in Fig. 1.

7 Conclusion

This paper proposed a Group Topic-Author model known as Latent Group Detective Model (LGDM) based on LDA to detect latent potential astroturfing groups (AGs). Astroturfing reviews and latent AGs are detected based on characteristics such as similarity in topic, context and time posted besides using sentiment found in the review. LGDM is based on the fact that there is difference between generative process of normal review and a potential astroturfing review. The LGDM is employed not only to discover AGs but also the sentiment exhibited by them. Towards this end AGs are analyzed with four cases. In the first case, AGs are analysed for similarity in topic, context and time posted. In the second case, AGs are analyzed based on the opposite sentiment. Case 2 AGs exhibit similarity in all aspects like content, context, time posted and sentiment. The third case throws light into high polarity sentiments associated with AGs. These AGs show similarity in content, context, time posted but have high polarity in sentiment. The fourth case is to analyze sentiments of AGs across products or services. The results are evaluated to know the efficiency of the LDGM in discovering latent AGs. We built a prototype application to demonstrate proof of the concept. Experimental results revealed the significance of LDGM as it performs better than other state of the art approaches. In our future work we improve sentiment analysis in LDGM with five point scale such as Highly Positive (HP), Positive (P), Neutral (U), Negative (N) and Highly Negative (HN) with different validation measures employed.

References

1. Akoglu, L., Chandy, R., Faloutsos, C.: Opinion fraud detection in online reviews by network effects. ICWSM **13**, 2–11 (2013)
2. Akoglu, L., Faloutsos, C.: Anomaly, event, and fraud detection in large network datasets. In: Proceedings of the Sixth ACM International Conference on Web Search and Data Mining, pp. 773–774. ACM (2013)
3. Cambria, E., Schuller, B., Xia, Y., Havasi, C.: New avenues in opinion mining and sentiment analysis. IEEE Intell. Syst. **28**(2), 15–21 (2013)
4. He, W., Zha, S., Li, L.: Social media competitive analysis and text mining: a case study in the pizza industry. Int. J. Inf. Manag. **33**(3), 464–472 (2013)
5. Hooi, B., et al.: BIRDNEST: Bayesian inference for ratings-fraud detection. In: Proceedings of the 2016 SIAM International Conference on Data Mining, pp. 495–503. SIAM (2016)
6. Mostafa, M.M.: More than words: social networks' text mining for consumer brand sentiments. Expert Syst. Appl. **40**(10), 4241–4251 (2013)
7. Mukherjee, A., Liu, B., Glance, N.: Spotting fake reviewer groups in consumer reviews. In: Proceedings of the 21st International Conference on World Wide Web, pp. 191–200. ACM (2012)
8. Mukherjee, A., Venkataraman, V., Liu, B., Glance, N.: Fake review detection: classification and analysis of real and pseudo reviews. Technical report, UIC-CS-2013-03, University of Illinois at Chicago (2013)
9. Nguyen, T.H., Shirai, K., Velcin, J.: Sentiment analysis on social media for stock movement prediction. Expert Syst. Appl. **42**(24), 9603–9611 (2015)
10. Rayana, S., Akoglu, L.: Collective opinion spam detection: bridging review networks and metadata. In: Proceedings of the 21th ACM SIGKDD International Conference on Knowledge Discovery and Data Mining, pp. 985–994. ACM (2015)
11. Savage, D., Zhang, X., Yu, X., Chou, P., Wang, Q.: Anomaly detection in online social networks. Soc. Netw. **39**, 62–70 (2014)
12. Song, J., Lee, S., Kim, J.: CrowdTarget: target-based detection of crowdturfing in online social networks. In Proceedings of the 22nd ACM SIGSAC Conference on Computer and Communications Security, pp. 793–804. ACM (2015)
13. Viswanath, B., et al.: Towards detecting anomalous user behavior in online social networks. In: USENIX Security Symposium, pp. 223–238 (2014)
14. Wang, H., Can, D., Kazemzadeh, A., Bar, F., Narayanan, S.: A system for real-time twitter sentiment analysis of 2012 us presidential election cycle. In: Proceedings of the ACL 2012 System Demonstrations, pp. 115–120. Association for Computational Linguistics (2012)

15. Wu, L., Hu, X., Morstatter, F., Liu, H.: Adaptive spammer detection with sparse group modeling. In: ICWSM, pp. 319–326 (2017)
16. Xiang, Z., Du, Q., Ma, Y., Fan, W.: A comparative analysis of major online review platforms: implications for social media analytics in hospitality and tourism. Tour. Manag. **58**, 51–65 (2017)
17. Xiao, C., Freeman, D.M., Hwa, T.: Detecting clusters of fake accounts in online social networks. In: Proceedings of the 8th ACM Workshop on Artificial Intelligence and Security, pp. 91–101. ACM (2015)
18. Zhao, J., Dong, L., Wu, J., Xu, K.: MoodLens: an emoticon-based sentiment analysis system for Chinese tweets. In: Proceedings of the 18th ACM SIGKDD International Conference on Knowledge Discovery and Data Mining, pp. 1528–1531. ACM (2012)

Visual Attention and Memory Augmented Activity Recognition and Behavioral Prediction

Nidhinandana Salian[✉]

Department of Computer Science Engineering, Manipal Institute of Technology,
Manipal, Karnataka, India
nidhisalian08@gmail.com

Abstract. Visual attention based on saliency and human behavior analysis are two areas of research that have garnered much interest in the last two decades and several recent developments have showed exceedingly promising results. In this paper, we review the evolution of systems for computational modeling of human visual attention and action recognition and hypothesize upon their correlation and combined applications. We attempt to systemically compare and contrast each category of models and investigate directions of research that have shown the most potential in tackling major challenges relevant to these tasks. We also present a spatiotemporal saliency detection network augmented with bidirectional Long Short Term Memory (LSTM) units for efficient activity localization and recognition that to the best of our knowledge, is the first of its kind. Finally, we conjecture upon a conceptual model of visual attention based networks for behavioral prediction in intelligent surveillance systems.

Keywords: Visual attention · Activity-recognition · Behavioral-prediction

1 Introduction

Intelligent visual surveillance in dynamic real-world scenes is an upcoming arena in the field of computer vision and automation and endeavors to recognize human activity by learning from prerecorded image sequences, and more generally to understand and predict subject behavior. A key precursor to arresting suspicious activity, which is the ultimate goal of most intelligent surveillance systems, is the accurate detection of peculiar or irregular movement that is observably deviant to the norm. Attention enhances the rationality of the agent by implementing a bottleneck which allows only relevant information to pass to higher level cognitive components like object recognition, scene interpretation, decision making and memory for further processing.

This paper attempts to bring dynamic visual saliency to the fore as a potential technique for anomalous activity discovery and surveys existing methodology that could provide a reliable means to recognize the most significantly noticeable features in videos of illegal activity and thus provide the basis for a predictive model to recognize the beginning of an occurrence of a potential crime. In our study, we have presented a complete overview of a variety of approaches taken to tackle the problem of saliency detection, and we suggest a biologically plausible framework applicable to the task of activity recognition and behavioral prediction in natural environments. Figure 1 is a diagrammatic representation of our model for saliency-based activity recognition.

Q. Chen et al. (Eds.): ATIS 2018, CCIS 950, pp. 92–106, 2018.
https://doi.org/10.1007/978-981-13-2907-4_8

Fig. 1. A brief overview of our model for saliency-based activity recognition

In order to understand dynamic saliency, it is important to explain the evolution of various lines of thought attempting to successfully model visual conspicuity in still images. The various approaches for saliency detection we have studied in our survey include the neuroscience-psychology based models that isolate prominent low-level features, task-specific models that serially scan a visual scene for target objects, reward-based reinforcement learning models, information maximization theory based on statistical rarity of local features and self-information, the computational-mathematical approach of regional covariance and the increasingly popular neural network based models. We have tried to present the advancement in this arena in chronological order as far as possible, presenting the pros and cons of each model and how subsequent developments attempt to meet the shortcomings of their predecessors.

The earliest models on behavioral prediction, on the other hand, were based on theoretic decision tree-like models of possible action sequences, assuming perfectly rational agents. Later models based on sample image sequences of human movement considered each agent capable of several internal states and introduced concepts such as mean gestures and gesture phases and used Dynamic Time Warping, Finite State Machines or Hidden Markov Models to explain transitions between them. More recent models introduced in the last two years have explored the idea of using various types of deep neural networks to learn predictive algorithms for human behavior with increasing accuracy. Most of these models use the given initial sequence of behavioral states or 'gestures' to output a predictive sequence of future states or gestures. Recent hypotheses have recently relabeled this task more appropriately as activity estimation, and the identification of the preparatory gestures as comparable to implicit activity recognition. Activity recognition is an important precursor to overall human behavioral analysis – because human behavior itself consists of several simultaneous activities. In our study, our primary focus will be on the activity recognition element of the existing paradigm, and how it could be integrated into visual attention prediction mechanisms for an end-to-end intelligent surveillance system for observable anomaly detection.

The remainder of this paper is structured as follows – Sect. 2(A) illustrates the origins and progression of computational architecture modelling visual attention, the shortcomings and successes of subsequent models. Section 2(B) presents a brief overview of the development and current dominant paradigms for behavioral prediction models and an introduction to the advancements in activity recognition models. In Sect. 3, we present a novel theory on a combined model of Saliency-Based Context-Aware Activity Recognition for implicit activity localization and identification using a pre-trained network for spatiotemporal saliency prediction, as a precursor to the conceptual model of Saliency based Behavioral Prediction, discussed in Sect. 4. We also address the limitations faced by the model from Sect. 3, which we are currently in the process of implementing, and the future work required for this approach to become a viable paradigm for intelligent surveillance systems.

2 Prior Work

2.1 Saliency Detection Models

Early theories attempting to explain visual saliency can be divided into two categories-top-down and bottom up, emulating the pre-attentive and attentive phases of attention in human cognition. While bottom up approaches were stimulus-driven and assumed that a spectrum of low level features in a scene were processed in parallel to produce cortical topographical maps, top-down features considered the approach that visual observation was goal-oriented and conducted by serially attending probable target locations. Koch and Ullman [1] were the first to recognize the similarity between the underlying neural circuitry and a hierarchy of elementary feature maps in the 80 s, inspired by the pioneering work of Treisman et al. on feature integration theory [2]. The first working model of this proposed architecture was introduced by Itti et al. [3]. It was later discovered that their winner-takes-all approach was neurally plausible and it was likely that there was a similar reward based mechanism used for memorization and learning within neurons in the brain [4, 5] and many reinforcement learning models were proposed in abidance with this line of thought [6, 7]. More research by neuro-biologists revealed a rapid decline in visual acuity while moving from the center of the retina(fovea) towards the periphery, and the need to rapidly realign the center of the eye with the region of focus by means of saccades [5, 8]. This provided a plausible explanation for the center bias observed in human eye fixation data over static images, and later models that took this into account proved far more effective than earlier architectures [9–12]. This theory was later advanced by the belief in the existence of a deictic system in the brain for memory representation in natural tasks, in order to economize the internal representation of visual scenes [5, 8, 13, 14]. Observations of individual eye movements in later experiments were found to suggest that information was incrementally acquired during a task and not at the beginning of a task [15]. Similar studies also showed that the initial internal representations most likely included illusory conjunctions formed between non-attended locations [2, 16] and was used to form a spatial 'gist' of the visual scene. This theory has recently been revisited by researchers attempting to model visual attention and memory [17].

Around the same period, the idea of a fast, automatic mapping of elementary features followed by serial scan of 'interesting' locations began to develop and many computational models around this time tried to replicate human eye fixations by using either objective, stimulus-driven or subjective, task-driven approaches [16, 18–20]. Thus the idea of selective visual attention was popularized and several decision-theoretic models based on sequential decision making, mostly emulating search tasks, were introduced [21]. Information theoretic models published at around the same time introduced the influence of prior experience and self-knowledge as a factor in gaze allocation [22, 23].

Advancements in neurobiological research around this time revealed the shared and specific neural substrates involved in the processing of perceptual identification (ventral stream) and spatial localization (dorsal stream) of target objects [24]. The conspicuity of spatially (pop out) and temporally (surprise motion) scarce factors was later thought to be an aspect of top down processing of visual scenes [25], through various research into inattentional blindness while performing a specific task [21]. Of late, a new notion that combines these two approaches has been suggested – introducing the idea of simultaneous covert and overt visual attention, stating that initially the brain creates a low level saliency mapping that takes into account the elementary features such as colour, orientation and intensity; and then conveniently uses that initial representation while pursuing a given task, oblivious to changes unless explicitly notified [26, 27]. Several prior experiments into this arena have shown results consistent with this theory, many recent models even relying upon advancements in virtual reality technology to create realistic yet completely controllable models of natural environments [16].

Although most of the existing neural network models are trained on vast eye fixation data [28–31] for static image viewing in controlled environments, even using neural networks to learn hidden non-linear mappings between images and fixations [32–34], new research has shown that this is a potentially problematic paradigm that presents a stark contrast to human eye fixations in natural, uncontrolled environments [35]. Recent studies have shown that although humans tend to fixate on faces and text irrespective of the task while watching images and videos in a controlled environment [14, 27, 36], this is not the case in real-world interactions, presumably because the faces in context belong to other human beings who could possibly look back at the observer [35, 37, 38], and similar experiments in the field of human psychology corroborate this finding, revealing that this potential for social interaction could possibly cause one to experience an elevated level of self-consciousness [39, 40]. This once again raises questions over the current state-of-the-art deep learning based models that use pre-trained object detection CNNs to fine tune saliency detection [41], on the assumption that humans predominantly tend to fixate on faces, text and objects [42, 43]. Multi-scale [44, 45] and multi-stream [46–50] convolutional networks and recurrent saliency detection models [34, 51] that have been introduced relatively recently, appear to provide an acceptable alternative that could perform equally well in controlled and natural environments.

2.2 Behavioral Prediction Models

Foreseeing the conduct of human members in strategic settings is a vital issue in numerous areas. Early work in this arena involved mostly expert-constructed decision theoretic models that either expected that members are superbly rational, or endeavored to show every member's intellectual forms in view of bits of knowledge from subjective brain research and experimental economics [52].

Behavioral prediction and gesture analysis models, on the other hand, have seen rapid progress in the models that consider a human to be a complex mechanical device, capable of several internal (mental) states that may be indirectly estimated. The prevalent notion in this area of research is that behavior can be divided into a combination of sequential, interleaving and concurrent activities, which in turn can be represented as a combination of actions. There are two main monitoring approaches for automatic human behavior and activity estimation, viz., vision-based [53, 54] and sensor-based [55, 56] monitoring. In this study, we will focus primarily on the vision-based behavioral approximation models.

The very first models that tried to capture the temporal aspects of movement tried to combine images from multiple viewpoints at multiple instants of time and tried to create a binary feature vector in the form of some variation of a temporal template [57, 58]. This traditional paradigm relied primarily on a collection of static image samples to calculate a mean gesture and then attempted to generalize this description to the rest of the data. The earliest models that set the precedent for prototype-based matching and tracking for gesture analysis that could clearly capture the smoothness and variability of human movement were based on Dynamic time Warping (DTW) [59] for temporal scale invariant action recognition and Finite State Machines (FSMs) [60], that considered gestural phases to be disjointed finite states, connected by transitions. Experimental results also converged with the theory that human actions are more suitably described as a sequence of control steps rather than as a combination of random positions and velocities. In order to decode states from observations, the idea of using Hidden Markov Models – a concept then being explored in text prediction and speech recognition models [61, 62], eventually made its way to behavioral analysis as well. In later models, prototypical dynamic movements that form the basic elements of cortical processing were represented as a large set of linear controllers (e.g. augmented Kalman filters), that could be sequenced together with a Markov network of probabilistic transitions to represent motion and behavior [63]. These were succeeded by models based on Conditional Random Fields (CRFs) [55, 64] that could capture long range dependencies better than traditional HMMs. This resulted in efficient models that could anticipate subject movements and output reasonably accurate predictive sequences that simulate typical subsequent human behavior from initial preparatory gesture analysis.

With the increasing availability of large amounts of annotated datasets and the advent of neural networks, many new models have been introduced for the purpose of behavior prediction, with alarmingly accurate results, thus creating a new paradigm for model-free, highly flexible and non-linear generalization [55, 65]. Time-dependent neural networks make use of a time-delay factor and use preceding values for future predictions [66, 67]. Self-organizing Kohonen networks could efficiently learn characteristics of normal trajectories and thus detect novel or anomalous behavior [68], comparable to earlier non-linear SVM models [69]. Game theoretic deep learning models and adversarial networks endeavor to simulate multi-agent strategic interactions [70]. More recently, sensor-based recurrent neural network models have been introduced that attempt to remember previously seen activity transitions in smart home environments and predict typical behavior from given initial state information [71].

Although these models did not explicitly specify activity recognition as an essential prerequisite to behavioral prediction, it has since been agreed upon that identification of the ongoing activity can in fact significantly improve the predictions on future behavioral patterns [72]. This pattern is particularly observable in state-based prediction models. Since our paper focuses on vision-based paradigms, we have restricted our review of action recognition systems to vision-based models. The general framework for action recognition system are similar to behavioral prediction models, with a low level processing that takes care of background subtraction, feature extraction, object detection and tracking, mid-level activity state estimation and a high level reasoning engine that outputs predictions. Approaches to activity recognition are either single-layered (space-time or sequential approaches) or hierarchical (statistical, syntactic or description-based approaches). Single layered space-time paradigms are essentially divided into feature-based (global or local activity representations) [57, 59, 73] that are commonly used in intelligent surveillance/tracking or smart home systems, or cognitive psychology research - inspired human body-model based (kinematics or 3D pose estimations) [74] that are primarily used for gesture recognition in gaming systems. The most noticeable limitation in the case of the latter class of models is that it is computationally expensive and prone to failure in cases of partial occlusion and multiple simultaneous activities. Our paper is in part inspired by the idea behind initial feature-based exploratory models that tried to capture activity as a collection of space-time interest points [73], but by way of neural networks we have overcome the many non-trivial encumbrances involved in the implementation of these early models, including the selection of optimal space time descriptors and clustering algorithms for minimally sized codebook construction.

Advancements in activity recognition have been most significantly noticeable in recent developments in video description models. These descriptive models use embedded information from previously seen training data like a visual vocabulary to optimize classification of new, unseen videos [75]. By incorporating memory modules [76] and recurrent loops [47] into the system, they have been able to incorporate the semantic and hierarchical nature of complex dynamic scenes and events into predictions. Although effective in appropriately constrained, uncluttered environments, the deterministic nature of these models makes them more susceptible to noise and occlusion that is inevitable in natural environments.

Fig. 2. Sample retrieval of spatio-temporally salient regions by our model in low resolution videos of concurrent activity.

3   Saliency Based Action Recognition Model

In this section, we present an LSTM-based model for robust and accurate activity recognition, augmented by a pre-trained spatio-temporal saliency detection network for the purpose of implicit activity localization and embedding of contextually significant information found in successive video frames. Human activity recognition is a vital field of interest in computer vision research. It has wide-ranging applications that include intelligent surveillance, patient monitoring systems and a variety of upcoming technology that incorporates human-computer interfaces into everyday devices.

In our model, we use visual attention to identify the most conspicuous distinctive characteristics of an activity followed by bi-directional Long Short Term Memory (LSTM) to remember and recognize these descriptive individualities in previously unseen videos, and output predictive labels classifying the ongoing activities. The Visual Attention module essentially acts as a spatial auto-encoder for perception and motion saliency.

Fig. 3. Sample STSHI output for a window of 30 frames that captures the action of opening a computer.

The Memory module, by taking as input the salient regions in consecutive frames, records temporal information and learns transitory semantics characteristic of each frame, while also retaining contextually relevant information that provides crucial cues about the environment. Theoretically, this approach should be feasible in uncontrolled, cluttered environments as well, since the prominent regularities that enable us to identify the occurrence of a particular activity, should still be perceptible to the visual attention network, and thus comprehensible to the memory module.

The saliency detection network we have used for our visual attention module is inspired by the model presented in this paper [68], and our current implementation of the Visual Attention Module makes use of the filters provided by OpenCV 3's Saliency module to provide a simple, functional end-to-end prototype for this purpose. Ideally, this would be replaced by a fine-tuned multi-stream convolutional neural network for spatio-temporal saliency detection. However, since most state of the art models are trained primarily on human eye fixation data, they work best on high resolution images of uncluttered environments, and their output is limited to a small number of highly salient points in the image, that cannot effectively capture important contextual information. Figure 2 is an example to show the effectiveness of our module even in very low resolution video frames and poor indoor lighting conditions. The weight matrix of the visual attention unit is then set to non-trainable while training the LSTM network for activity recognition.

Since the input to the visual attention network is in video format while the input to the memory module is in the format of a series of images, it is important to reduce the temporal aspect of movement in the three-dimensional(X, Y, T) input data into a two-dimensional(X, Y) representation and this is done by concatenating a small window of successive saliency predictions along the time axis.

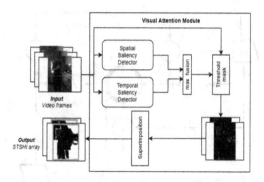

Fig. 4. Overview of the visual attention module.

Superimposition of only the temporally salient regions indicative of motion, extracted using Dense Optic Flow detection, while retaining common spatially salient characteristics, allows us an optimal representation of micro-gestures within a brief temporal window. The input to the memory module is thus similar to Motion History Images [59] or Spatio-Temporal Shape Templates [73], essentially consistent of concatenated blobs of Salient regions in consecutive video frames and shall hereinafter be referred to as Spatio-Temporal Saliency History Images (STSHI) instead. The 3-channel data is finally reduced to a single grayscale thresh map representation. Figure 3 is an example for an STSHI template formed by concatenating features from a window of 30 consecutive video frames, and clearly represents the simple action of a laptop computer being opened. An array of STSHI templates corresponding to a single video is then fed to the bi-directional LSTM network that intrinsically learns a mapping function to match the input STSHI templates to the correct activity category. Figure 4 illustrates the functioning of the Visual Attention Module, and the intermediate representation of the salient regions before superimposition. We have also chosen to maintain the original dimensions of the input frames, so the network may also learn from the relative spatial orientation of salient features, a characteristic from which the mind would immediately derive a gist of the visual scene.

Our memory model was inspired by the idea behind earlier models that, despite having implemented neither visual saliency nor long term dependencies, attempted to 'match' videos using spatio-temporal feature relationships [77]. The most recent models of Long Short Term Memory cells [78], have been shown to consistently outperform standard Recurrent Neural Networks in learning tasks that have involved identification of persistent, long term dependencies. Figure 5 is a simplified depiction of the internal functioning of an LSTM cell. Figure 6 describes the equations for the learning process of an LSTM cell at a given time step t.

A recurrent neural network, consistent of two serially connected bi-directional LSTM units has been proposed for our memory module instead of a standard LSTM because this model is intended to be a precursor to a behavior analysis model, and it is useful to simultaneously learn encoding and decoding, from saliency templates to activity labels, and vice versa.

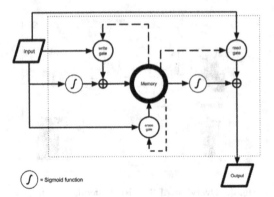

Fig. 5. Components of an LSTM cell

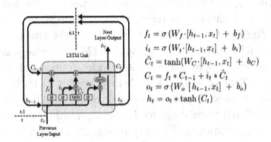

Fig. 6. Learning process in an LSTM cell

It also allows for some amount of variation in the sequence of primitive gestures that make up an activity and we conjecture that it makes the model invariant to a measurable amount of temporal scaling. During the fully supervised training process, each input to the memory module is essentially the output obtained from the primary module for visual attention and consists of an array of resized STSHI to represent the original input video. Resizing is done in order to reduce the number of features being fed to the network, but care is taken to maintain the original height-to-width ratio so as not to distort the saliency detections. The target output is a binary HOT-encoded vector of length equivalent to total number of possible activity classes, with only the elements corresponding to activities depicted in the video set to 1.

The first bi-directional LSTM unit takes the input and produces one output per template in an STSHI sequence, using the TimeDistributed wrapper layer in Keras. Since each template essentially captures a micro-gesture in the activity, the first memory unit effectively learns to recognize micro-gestures and then sends a sequence of outputs to the second memory unit. The second bi-directional LSTM unit also utilizes a TimeDistributed wrapper layer, and condenses the series of outputs into a single output vector. This thus reduces the first layer's outputs, which corresponded to each constituent gesture in an activity video, into a single output corresponding to the overall activity being performed in the video. A dense layer is then used to further format the vector into the shape of the expected output vector. The final output of the memory module is a vector of the same length as the target output vector, and each element of the vector is a probability score denoting the likelihood of the corresponding activity being performed in the input video. By setting an optimum threshold limit upon these confidence predictions, we can successfully eliminate the consideration of unlikely activities, while preserving predictions for possibly concurrent activities in the same video. The easiest way to do this would possibly be to consider as the threshold a percentage (e.g.: 80%) of the highest confidence score outputted, provided that the maximum is high enough to indicate certain presence of the corresponding activity.

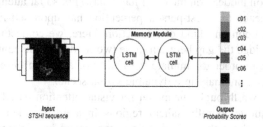

Fig. 7. High-level overview of the memory module

The dataset we will be using for training and testing purposes is the publicly available Charades dataset of daily human activities in indoor environments [79], scaled up to 480p. The videos consist of one or two human subjects performing common activities such as writing, walking, using a computer etc. Figure 7 illustrates the architecture of our Memory Module.

Our entire model can be considered an analogy of the human anatomy responsible for the processing of visual information and therein lies its inherent neurobiological plausibility. The visual attention module's recognition of spatially and temporally significant locations is comparable to the splitting of motion and detail visual information, i.e., the magno and parvo cellular pathways into the cortical area (V1), via the lateral geniculate nucleus (LGN) of the thalamus. Parvocellular ganglion retinal cells being smaller and slower, carry many details such as colour, intensity etc. Magnocellular ganglion retinal cells being larger, faster and rather rough in their representations, carry transient information such as motion details. Since we only retain the most salient regions for the STSHI templates, we also incorporate the aspect of foveated vision at interesting locations. We do not deploy background subtraction as in many popular convolutional models, because activity recognition is highly contextual, and studies have proven that the brain unconsciously utilizes the awareness of several spatial cues to form illusory conjunctions about the surrounding environment. The memory module models the short term memory of humans and the tendency to recognize the reoccurrence of prior experiences by retaining an internal representation of their distinctive temporal semantics. As such, our model is a better prototype for human activity recognition in real world environments as opposed to activity recognition in controlled virtual simulations or video viewing experiments.

4 Saliency-Based Behavioral Prediction – Discussion

Experiments have shown that human behavior is the outcome of interacting determinants and that every individual action is an aggregate of self-intention, social interaction and contextual environment. It is evident from our study that all the developments in saliency detection and behavioral analysis up until now have not completely considered critical factors in the estimation and prediction of a human entity's natural gaze or motion. Saliency detection frameworks have so far failed to account for important aspects of cognitive psychology and behavior that, as prior self-information, are imperative to understanding visual perception in the real world. Behavioral prediction models, on the other hand, have not so far attempted to look into the inevitable influence that first-person perception has upon self-information accumulation and thus, upon any kind of interaction. Here, we conjecture upon a model with the intent to bridge the gap between these two models in the context of intelligent visual surveillance, in order to simulate the ability of a human observer to detect an observable behavioral anomaly in a dynamic visual scene.

As we have shown through our model for visual attention-based activity recognition, detecting spatio-temporally salient regions in a given video can significantly improve predictions about the activity being represented in it. Since our model is a single layered space-time based approach to activity recognition, it is clear that advancing it to incorporate the hierarchical nature of behaviors that are composite of atomic activities, is probably the best possible way to analyze and predict subsequent behavior. Previously, hierarchical models for this purpose adopted symbolic artificial intelligence techniques and were primarily based on statistical (Bayesian Networks [80] and Markov Models [63]) or syntactic (Stochastic Context Free Grammars [81]) approaches, but recent developments in deep learning – specifically recurrent neural networks [71], have significantly outperformed earlier models, and appear to provide a more viable alternative for probabilistic predictive analysis of human behavior. Since all behavioral analysis models ultimately attempt to capture the transitions between internal states within the human brain by way of understanding observable external information, it makes sense that the solution would probably lie within a combination of various neural networks for learning and memory – a concept that is itself based on recreating neuronal interconnections within the brain.

Distinctive behavior could easily be described as individual mannerisms applied to common performative actions. Our proposed model intends to exploit this property by trying to create a relatively change invariant activity recognition model, so as to be reasonably robust to personal affectations. Thus we can generalize behavioral tendencies based on context and develop a system to predict future actions and detect anomalous or suspicious movements. In future adaptations of our proposed model, we intend to introduce an additional recurrent network architecture, to learn sequences of the activity labels predicted by our current system, and thus classify overall behaviors. This system could then be applied in scenarios where intelligent surveillance of dynamic environments is required, to detect atypical behavior patterns.

References

1. Koch, C., Ullman, S.: Shifts in selective visual attention: towards the underlying neural circuitry. Hum. Neurobiol. **4**(4), 219–227 (1985)
2. Treisman, A., Gelade, G.: A feature-integration theory of attention. Cogn. Psychol. **12**(1), 97–136 (1980)
3. Itti, L., Koch, C., Niebur, E.: A model of saliency-based visual attention for rapid scene analysis. IEEE Trans. Pattern Anal. Mach. Intell. **20**(11), 1254–1259 (1998)
4. Ballard, D.H., Hayhoe, M., Pelz, J.: Memory representations in natural tasks. J. Cogn. Neurosci. **7**, 66–80 (1995)
5. Ungerleider, S.: Mechanisms of visual attention in the human cortex. Annu. Rev. Neurosci. (2000)
6. Itti, L., Koch, C.: A saliency-based search mechanism for overt and covert shifts of visual attention. Vis. Res. **40**, 1489–1506 (2000)
7. Minut, S., Mahadevan, S.: A reinforcement learning model of selective visual attention. In: Autonomous Agents Conference (2001). Author, F.: Article title. Journal **2**(5), 99–110 (2016)
8. Hillstrom, A.P., Yantis, S.: Visual-motion and attentional capture. Percept. Psychophys. **55**, 399–411 (1994)
9. Hamker, F., Worcester, J.: Object detection in natural scenes by feedback. In: Biologically Motivated Computer Vision: Second International Workshop, pp. 398–407 (2002)
10. Jovancevic, J., Sullivan, B., Hayhoe, M.: Control of attention and gaze in complex environments. J. Vis. **6**, 1431–1450 (2006)
11. Tatler, B.W.: The central fixation bias in scene viewing: selecting an optimal viewing position independently of motor biases and image feature distributions. J. Vis. **7**(14), 1–17 (2007)
12. Hou, X., Zhang, L.: Saliency detection: a spectral residual approach. In: CVPR 2007. IEEE (2007)
13. Goodale, M.A., Milner, A.D.: Separate visual pathways for perception and action. Trends Neurosci. **15**(1), 20–25 (1992)
14. Cerf, M., Frady, E., Koch, C.: Faces and text attract gaze independent of the task: experimental data and computer model. J. Vis. **9**, 10.1–1015 (2009). https://doi.org/10.1167/9.12.10
15. Jovancevic-Misic, J., Hayhoe, M.: Adaptive gaze control in natural environments. J. Neurosci. **29**(19), 6234–6238 (2009). https://doi.org/10.1523/JNEUROSCI.5570-08.2009
16. Folk, C.L., Remington, R.W., Johnston, J.C.: Involuntary covert orienting is contingent on attentional control settings. J. Exp. Psychol. **18**(4), 1030–1044 (1992)
17. Wu, C., Wang, H., Pomplun, M.: The roles of scene gist and spatial dependency among objects in the semantic guidance of attention in real-world scenes. Vis. Res. **105**, 10–20 (2014)
18. Shinoda, M., Hayhoe, M.M., Shrivastava, A.: What controls attention in natural environments. Vis. Res. **41**, 3535–3545 (2001)
19. Triesch, J., Ballard, D.H., Hayhoe, M.M., Sullivan, B.T.: What you see is what you need. J. Vis. **3**, 9 (2003)
20. Oliva, A., Torralba, A., Castelhano, M., Henderson, J.: Top–down control of visual attention in object detection. In: Proceedings of the 2003 International Conference on Image Processing (2003)
21. Bruce, N., Tsotsos, J.: Saliency, attention, and visual search: an information theoretic approach. J. Vis. **9**, 5 (2009)

22. Lee, T.S., Stella, X.Y.: An information-theoretic framework for understanding saccadic eye movements. In: NIPS (1999)
23. Bruce, N., Tsotsos, J.: Saliency based on information maximization. In: Advances in Neural Information Processing Systems (2006)
24. Fink, G.R., Dolan, R.J., Halligan, P.W., Marshall, J.C., Frith, C.D.: Space-based and object-based visual attention: shared and specific neural domains. Brain **120**, 2013–2028 (1997)
25. Itti, L., Baldi, P.: A principled approach to detecting surprising events in video. In: Proceedings of IEEE CVPR (2005)
26. Judd, T., Ehinger, K., Durand, F., Torralba, A.: Learning to predict where humans look. IEEE (2009)
27. Judd, T., Durand, F., Torralba, A.: Fixations on low resolution images. J. Vis. **11**(4), 14 (2011)
28. Jiang, M., Huang, S., Duan, J., Zhao, Q.: SALICON: saliency in context. In: CVPR 2015. IEEE (2015)
29. Jiang, M., Boix, X., Roig, G., Xu, J., Van Gool, L., Zhao, Q.: Learning to predict sequences of human visual fixations. IEEE Trans. Neural Netw. Learn. Syst. **27**(6), 1241–1252 (2016)
30. Cornia, M., Baraldi, L., Serra, G., Cucchiara, R.: Predicting human eye fixations via an LSTM-based saliency attention model. arXiv preprint arXiv:1611.09571 (2017)
31. Bylinskii, Z., et al.: MIT Saliency Benchmark (2017). http://saliency.mit.edu
32. Liu, Z., Zhang, X., Luo, S., Le Meur, O.: Superpixel-based spatiotemporal saliency detection. IEEE Trans. Circ. Syst. Video Technol. **24**, 1522–1540 (2014)
33. Kruthiventi, S.S., Gudisa, V., Dholakiya, J.H., Venkatesh Babu, R.: Saliency unified: a deep architecture for simultaneous eye fixation prediction and salient object segmentation. In: CPVR (2016)
34. Dodge, S., Karam, L.: Visual saliency prediction using a mixture of deep neural networks. arXiv preprint arXiv:1702.00372 (2017)
35. Tatler, B.W., Hayhoe, M.M., Land, M.F., Ballard, D.H.: Eye guidance in natural vision: reinterpreting salience. J. Vis. **11**(5), 5 (2011)
36. Kümmerer, M., et al.: Understanding low- and high-level contributions to fixation prediction. In: 2017 IEEE International Conference on Computer Vision (ICCV), pp. 4799–4808 (2017)
37. Foulsham, T., Walker, E., Kingstone, A.: The where, what and when of gaze allocation in the lab and the natural environment. Vis. Res. **51**, 1920–1931 (2011)
38. Tatler, B.W.: Eye movements from laboratory to life. In: Horsley, M., Eliot, M., Knight, B., Reilly, R. (eds.) Current Trends in Eye Tracking Research, pp. 17–35. Springer, Cham (2014). https://doi.org/10.1007/978-3-319-02868-2_2
39. Laidlawa, K.E.W., Foulshamb, T., Kuhnc, G., Kingstone, A.: Potential social interactions are important to social attention. PNAS **108**, 5548–5553 (2011)
40. Gobel, M.S., Kim, H.S., Richardson, D.C.: The dual function of social gaze. Cognition **136**, 359–364 (2015)
41. Murabito, F., Spampinato, C., Palazzo, S., Giordano, D., Pogorelov, K., Riegler, M.: Top-down saliency detection driven by visual classification. Comput. Vis. Image Underst. (2018)
42. Kruthiventi, S.S., Ayush, K., Babu, R.V.: DeepFix: a fully convolutional neural network for predicting human eye fixations. arXiv preprint arXiv:1510.02927 (2015)
43. Kümmerer, M., Wallis, T.S., Bethge, M.: DeepGaze II: reading fixations from deep features trained on object recognition. arXiv preprint arXiv:1610.01563 (2016)
44. Cornia, M., Baraldi, L., Serra, G., Cucchiara, R.: A deep multilevel network for saliency prediction. In: Proceedings of the International Conference on Pattern Recognition (2016)
45. Wang, W., Shen, J.: Deep visual attention prediction. arXiv preprint arXiv:1705.02544 (2018)

46. Simonyan, K., Zisserman, A.: Two-stream convolutional networks for action recognition in videos. In: NIPS, pp. 568–576 (2014)
47. Donahue, J., et al.: Long-term recurrent convolutional networks for visual recognition and description. In: CVPR (2015)
48. Pan, J., Sayrol, E., Giro-i Nieto, X., McGuinness, K., O'Connor, N.E.: Shallow and deep convolutional networks for saliency prediction. In: CVPR (2016)
49. Bak, C., Erdem, E., Erdem, A.: Two-stream convolutional neural networks for dynamic saliency prediction. arXiv preprint arXiv:1607.04730 (2016)
50. Bak, C., Kocak, A., Erdem, E., Erdem, A.: Spatio-temporal networks for dynamic saliency prediction. arXiv preprint arXiv:1607.04730v2 (2017)
51. Kuen, J., Wang, Z. Wang, G.: Recurrent attentional networks for saliency detection. In: CVPR (2016)
52. Unzicker, A., Juttner, M., Rentschler, I.: Similarity-based models of human visual recognition. Vis. Res. **38**, 2289–2305 (1998)
53. Hu, W., Tan, T., Wang, L., Maybank, S.: A survey on visual surveillance of object motion and behaviors. IEEE Trans. Syst. Man Cybern. **34**(3), 334–352 (2004)
54. Poppe, R.: A survey on vision-based human action recognition. Image Vis. Comput. **28**(6), 976–990 (2010)
55. Van Kasteren, T., Noulas, A., Englebienne, G., Kröse, B.: Accurate activity recognition in a home setting. In: Proceedings of the 10th International Conference on Ubiquitous Computing, pp. 1–9 (2008)
56. Avci, U., Passerini, A.: A fully unsupervised approach to activity discovery. In: Salah, A.A., Hung, H., Aran, O., Gunes, H. (eds.) HBU 2013. LNCS, vol. 8212, pp. 77–88. Springer, Cham (2013). https://doi.org/10.1007/978-3-319-02714-2_7
57. Bobick, A., Davis, J.: Real-time recognition of activity using temporal templates. In: Proceedings 3rd IEEE Workshop on Applications of Computer Vision 1996, WACV 1996, pp. 39–42 (1996)
58. Wilson, A.D., Bobick, A.F., Cassell, J.: Temporal classification of natural gesture and application to video coding. In: Proceedings of IEEE Conference on Computer Vision and Pattern Recognition, pp. 948–954 (1997)
59. Bobick, A., Davis, J.: An appearance-based representation of action. In: ICPR (1996)
60. Bobick, A.F., Wilson, A.D.: A state-based technique to the representation and recognition of gesture. IEEE Trans. Pattern Anal. Mach. Intell. **19**, 1325–1337 (1997)
61. Rabiner, L.: "A tutorial on hidden Markov models and selected applications in speech recognition", Proceedings of the IEEE, 1989
62. Fossler-Lussier, E.: Markov models and hidden Markov models: a brief tutorial. International Computer Science Institute (1998)
63. Pentland, A., Liu, A.: Modeling and prediction of human behavior. Neural Comput. **11**(1), 229–242 (1999)
64. Kim, E., Helal, S., Cook, D.: Human activity recognition and pattern discovery. IEEE Pervasive Comput. **9**, 48–53 (2010)
65. Phan, N., Dou, D., Piniewski, B., Kil, D.: Social restricted Boltzmann machine: human behavior prediction in health social networks. In: ASONAM 2015 (2015)
66. Yang. M., Ahuja, N.: Extraction and classification of visual motion pattern recognition. In: Proceedings of IEEE Conference on Computer Vision and Pattern Recognition, pp. 892–897 (1998)
67. Meier, U., Stiefelhagen, R., Yang, J., Waibel, A.: Toward unrestricted lip reading. Int. J. Pattern Recogn. Artif. Intell. **14**(5), 571–585 (2000)
68. Owens, J., Hunter, A.: Application of the self-organizing map to trajectory classification. In: Proceedings of IEEE Int. Workshop Visual Surveillance (2000)

69. Zhao, H., Liu, Z.: Human action recognition based on non-linear SVM decision tree. J. Comput. Inf. Syst. **7**(7), 2461–2468 (2011)

70. Hartford, J., Wright, J.R. Leyton-Brown, K.: Deep learning for human strategic behavior prediction. In: NIPS (2016)

71. Almeida, A., Azkune, G., Predicting human behavior with recurrent neural networks. In: Appl. Sci. 2018 (2018)

72. Sigurdsson, G., Russakovsky, O., Gupta, A.: What actions are needed for understanding human actions in videos? In: ICCV 2017, pp. 2156–2165 (2017)

73. Bregonzio, M., Gong, S. Xiang, T.: Recognising action as clouds of space-time interest points. In: Proceedings of International Conference on Computer Vision and Pattern Recognition (CVPR) (2009)

74. An, N., Sun, S., Zhao, X., Hou, Z.: Remember like humans: visual tracking with cognitive psychological memory model. Int. J. Adv. Robot. Syst. **14**, 1–9 (2017)

75. Plummer, B.A., Brown, M., Lazebnik, S.: Enhancing video summarization via vision-language embedding. In: CPVR (2017)

76. Fakoor, R., Mohamed, A., Mitchell, M., Kang, S.B., Kohli, P.: Memory-augmented attention modelling for videos. arXiv preprint arXiv:1611.02261v4 (2017)

77. Ryoo, M.S., Aggarwal, J.K.: Spatio-temporal relationship match: video structure comparison for recognition of complex human activities. In: Computer Vision, pp. 1593–1600 (2009)

78. Hochreiter, S., Schmidhuber, J.: Long short-term memory. Neural Comput. **9**(8), 1735–1780 (1997)

79. Sigurdsson, G.A., Varol, G., Wang, X., Farhadi, A., Laptev, I., Gupta, A.: Hollywood in homes: crowdsourcing data collection for activity understanding. In: Leibe, B., Matas, J., Sebe, N., Welling, M. (eds.) ECCV 2016. LNCS, vol. 9905, pp. 510–526. Springer, Cham (2016). https://doi.org/10.1007/978-3-319-46448-0_31

80. Oliver, N., Rosario, B., Pentland, A.: A Bayesian computer vision system for modeling human interactions. ICVS 1999. LNCS, vol. 1542, pp. 255–272. Springer, Heidelberg (1999). https://doi.org/10.1007/3-540-49256-9_16

81. Ryoo, M.S., Aggarwal, J.K.: Semantic representation and recognition of continued and recursive human activities. Int. J. Comput. Vis. **82**, 1–24 (2009)

Security Implementations

Distributed Computing Security Model Based on Type System

Yong Huang[1], Yongsheng Li[2(✉)], and Qinggeng Jin[1]

[1] College of Software and Information Security,
Guangxi University for Nationalities, Nanning 530006, China
[2] College of Information Science and Engineering,
Guangxi University for Nationalities, Nanning 530006, China
guxnsubmission@163.com

Abstract. Aiming at the problem of access control in distributed computing security model, a secure Seal calculus based on hybrid type detection is proposed. In order to realize the security policy that low security level information in the multi-level security system can only flow to equal or higher security levels, it establishes a security system with security level for Seal calculus, and uses static type detection to realize fine particle size access control efficiently. To solve the practical problems of channel control and security power reduction, an effective dynamic transformation framework based on mandatory type transformation is proposed. The static detection and dynamic detection are organically integrated to form a unified security model, which can not only guarantee the security of distribute computing, but also have good availability.

Keywords: Distributed computing · Seal calculus · Type system

1 Introduction

With the emergence and extensive application of new distributed computing technology such as autonomous computing, mobile computing and cloud computing, the implementation of multi-level security protection in complex distributed environment has become an important challenge in the field of security research [1, 2]. The main reason for the emergence of multi-level security policies is to prevent unauthorized disclosure of threats among different levels of security information. Therefore, there are inherent multi-level security needs in both the military and the commercial fields.

In recent years, the mobile Internet as the representative of the new distributed computing, has achieved some results in the study of security features. Security work mainly includes two categories: (1) For mobile terminals, access networks and application services and application of basic research, specifically including mobile terminal security, wireless network security, application security, content security and location privacy protection; (2) Based on the mobile computing mobile Internet mechanism research, including mobile computing security model, mobile security assessment and so forth. In this paper, we focus on the research work of (2). In terms of the mechanism modeling of mobile computing, the π calculus and Seal calculus are widely used in mobile computing due to the effective creation and exchange of channel names. As described in [3], the π calculus is further extended to distinguish between two different capability types: the read channel type and the write channel type, the subtype relation is established, and the algebraic theory of the typed moving process is given. Similarly, [4–6] gives the Seal calculus type system, distinguishes the message type and the Agent type, establishes the subtype relation, and then gives the algebraic theory of the Seal calculus type system. The literature [7] uses the local position and the different positions of the computing power to describe the mobile system, and then use the tag Kripke structure to establish its formal semantics in the form of network and use the SATABS model checker to verify. However, [3–7] do not involve system security features. In order to analyze the security in mobile computing, the literature [8] for security level division uses π calculus modeling mobile resource access to ensure that subjects with low security levels can't access high security level information. Obviously, this does not guarantee confidentiality and integrity at the same time. In view of the shortcomings of the literature [8], Guo et al. [9] based on the π calculus type system gives a security model which guarantees the confidentiality and integrity at the same time.

However, the above research does not take into account the impact of location on system security, or the limited expression capacity of formal tools which can not explicitly describe the system's distribution characteristics, location concepts and combination characteristics. Therefore, we will model and analyze the distributed system by means of the calculation model tool-Seal, which can clearly describe the location nesting, program movement and resource access control characteristics. By establishing a secure Seal type system, the access control is abstracted as I/O operation (read and write operations) of a distributed system, in the program language level to achieve fine-grained access control. In order to solve the problem of channel control and safety power reduction in distributed system, on the basis of Seal type system, we propose an effective dynamic transformation framework based on mandatory type transformation, which integrates static detection and dynamic detection to form a unified distributed system security form model.

2 Security Seal Calculus Type System

In this section, we will draw on the types of programming language features and multi-level security definition to introduce the concept of security level in the type system and propose a security Seal calculus for hybrid type detection. Security Seal calculus is a type system of Seal calculus. We can guarantee that high security level information does not flow to low security level by static type detection. We propose a dynamic transformation framework based on forced type transformation for static power reduction to realize the organic combination of static detection and dynamic detection and form a unified dynamic security form model. It should be emphasized that although this article is based on a two-level security model description, but the definition of the security framework is easy to promote the application to multi-level security distributed system.

2.1 Security Type

This section defines the complete lattice of the security level $(L, \preceq, Top, Bot)$, which is easy to generalize to any level, that is, not limited to "high" and "low", and L is the security level set. We use $\sigma, \beta, \cdots$ to indicate the level of security and use Top and bot to represent the highest security level and the minimum security level, respectively, and there is $Bot \preceq Top$. We use lev (x) to denote the security level of x, where x is any name or type and define the type with the security level as follows:

Definition 2.1.1 (Security Type)

$$V ::= M | A$$
$$M ::= (B, \sigma)$$
$$A ::= \{x_1 : V_1, \cdots, x_n : V_n\}^r | \{x_1 : V_1, \cdots, x_n : V_n\}^w | \{x_1 : V_1, \cdots, x_n : V_n\}^{rw}$$

By definition we can see that the security type V includes the message type M (the type of security used to describe the variable, channel name, seal name, or information) and interface type A (the type of security that describes the interface capability). Message type M is a two-tuple (B, σ), where the first dimension can represent different types of fields, and in addition to describing the common channel names and seal names, it can also represent a set of security categories for messages, such as legal messages {CN, HK}; The second dimension indicates the security level of the message, corresponding to the public, confidential, and top secret in the actual application. In reality, B and σ will be different with different application environment, so we do not explicitly limit the physical meaning of B and σ. The interface type A is represented by a set, and as described in [10], the interface can also be regarded as Agent or Seal, which is a set of channels. If each element in the interface set is a read interface type $\{x_1 : V_1, \cdots, x_n : V_n\}^r$, it is not formalized to represent the interface set in A^r; if each element in the interface set is a write interface type $\{x_1 : V_1, \cdots, x_n : V_n\}^w$, it is not formalized to represent the interface set in A^w; If the elements in the interface set have both write and read interface types, they are not formalized to represent the interface set A^{rw}. If the variable x is of type V^r, it means that the value of type V can be read into channel x with a security level of $lev(V)$. Similarly, if the variable x is of type V^w, this means that a value of type V can be written to channel x with a security level of $lev(V)$. Obviously, the security level of the interface type $lev(V)$ indicates the security level of the data through the interface, which reflects a certain capability of the system. According to the multi-level security features, we can know that the information with level σ can only flow to lower (or the same) security level information. Therefore, from the point of view of the access control mechanism, the definition of the security type embodies the most intuitive security policy: high security level information can't be read and written through the low security level channel.

2.2 Subtype Rules

To define the subtype rules on the security type, we first discuss the relationship of the message security type pairs on each dimension. For the first dimension, $\leq$ is used to denote the subtype relation on the type field. The definition of the relationship is different depending on the different cases. If it is a numeric type, the comparison operation can be performed according to the algebraic rule. If it is a set type, the subtype relation $\leq$ on the message security type field is $\{HK\} \subseteq \{CA, HK\}$, which is equivalent to the inclusion $\subseteq$ on the set, as the message security type field CA and HK in the previous section. For the second dimension, as described in the previous section, Information flow with security level can be regarded as a complete grid. Therefore, in the security level it constitutes a partial order relationship (with the expression $\preceq$), its meaning is as follows: If σ_1 and σ_2 are two elements on the security label collection, $\sigma_1 \preceq \sigma_2$ indicates that the security identified as σ_1 is less than the security identified as σ_2. Thus, similar to the literature [6], the subtype rules on the security type can be defined as follows:

$$SUB - REFL \qquad \frac{SUB - TRANS}{\frac{S \leq S', S' \leq T}{S \leq T}}$$
$$\overline{T \leq T}$$

$$SUB - DATA1 \qquad SUB - DATA2$$
$$\frac{\sigma_1 \preceq \sigma_2}{(B, \sigma_1) \leq (B, \sigma_2)} \qquad \frac{B_1 \leq B_2}{(B_1, \sigma) \leq (B_2, \sigma)}$$

$$SUB - DATA3 \qquad SUB - WRITE$$
$$\frac{B_1 \leq B_2, \sigma_1 \preceq \sigma_2}{(B_1, \sigma_1) \leq (B_2, \sigma_2)} \qquad \frac{V_1 \leq V_2}{A_2^w \leq A_1^w}$$

$$SUB - READ - I \qquad SUB - LINK$$
$$\frac{V_1 \leq V_2}{A_1^r \leq A_2^r} \qquad \frac{(\forall j)(\exists i) V_i \leq V_j'}{\{V_i\}_{i \in I} \leq \{V_j\}_{j \in J}}$$

$$SUB - READ - II \quad SUB - RW - R \quad SUB - RW - W$$
$$\frac{\sigma_1 \preceq \sigma_2}{(A^r, \sigma_1) \leq (A^r, \sigma_2)} \qquad A^{rw} \leq A^r \qquad A^{rw} \leq A^w$$

Sub-type rules defined by the above table, the meaning is as follows: Rules *SUB-REFL* and *SUB-TRANS* indicate that $\leq$ is a preorder relationship, that is, the security type meets the reflexive, transmission, asymmetry. Rule *SUB-DATA1* displays: If σ_2

has a higher security level than σ_1, the message type (B, σ_1) is a subtype of message type (B, σ_2). Rule *SUB-DATA2* indicates that if the type field B_1 is a subtype of B_2, the message type (B_1, σ) is a subtype of the message type (B_2, σ). Rule *SUB-DATA3* indicates that if σ_2 has a higher security level than σ_1 and the type field B_1 is a subtype of B_2, the message type (B_1, σ_1) is a subtype of the message type (B, σ_2). Rule *SUB-WRITE* indicates that if the message type V_1 is a subtype of V_2, then the write interface type A_2^w is a subtype of A_1^w. The rule *SUB-READ-I* indicates that if the message type V_1 is a subtype of V_2, then the read interface type A_1^r is a subtype of A_2^r. The rule *SUB-READ-II* indicates that if the security level of σ_2 is higher than σ_1, then the read interface type (A^r, σ_1) is a subtype of (A^r, σ_2). That reflects the subtype relationship of the input types between the different security classes. Its intuitive meaning is that the low-security level of the channel read data can also be read at a high security level channel. Rules *SUB-RW-R* and *SUB-RW-W* indicate that the types of read-write interfaces can construct types that can read and write interfaces.

2.3 Type Rules

In the literature [6], we define the form of the type: $\Gamma \vdash *: \mathsf{T}$, where Γ denotes the type environment, * denotes the process, name, agent or seal, T denotes the type, The whole formula shows that in the environment of Γ, the type of * is T. Here we give the security Seal typing rules:

As noted in the literatures, subtle changes in the rules of the type system will have a significant effect on the nature of the whole system, so we need to prove the relevant properties. Among them, Subject Reduction and Subject Congruence are two very important properties of the type detection. Here, the definition of the subject protocol is: if the type of system meets the Subject Reduction, then the results of each step of the good type process are still good type. The security nature of the security seal reflected in the security Seal calculation is that if a process satisfies a certain security level in the initial state, it will remain below a certain security level after each operation. The definition of the Subject Congruence is that if the type system satisfies the Subject Congruence, the process of the good type process congruence is still good type. The security nature of a security Seal type system is that if a process is of some security type, the equilateral process also satisfies a certain type of security. In the initial state, the security of the system is provided by the type definition, so that if the good type process meets Subject Reduction and Subject Congruence, the system would be able to maintain security.

$$T-name \quad \frac{}{\Gamma_c \vdash x : T} \qquad T-base \quad \frac{bv \in B}{\Gamma_c \vdash bv : B}$$

$$T-Interface \quad \frac{\forall i = 1, \cdots, n \; \Gamma_c \vdash x_i : V_i}{\Gamma_c \vdash [x_1 : V_1, \cdots, x_n : V_n]}$$

$$T-par \quad \frac{\Gamma_c \vdash P_1 : A \quad \Gamma_c \vdash P_2 : A}{\Gamma_c \vdash P_1 \mid P_2 : A} \qquad T-rep \quad \frac{\Gamma_c \vdash P : A}{\Gamma_c \vdash ! P : A}$$

$$T-res \quad \frac{\Gamma, x : M_c \vdash P : A}{\Gamma_c \vdash (vx : M) P : A} \; x \notin fn(A)$$

$$T-rd-local \quad \frac{\Gamma_c \vdash x : M \quad \Gamma, y : M_c \vdash P : A}{\Gamma_c \vdash x^* \langle y : M \rangle P : A}$$

$$T-rd-up \quad \frac{\Gamma_c \vdash x : M \quad \Gamma, y : M_c \vdash P : A}{\Gamma_c \vdash x^\uparrow \langle y : M \rangle P : (A \oplus [x : M])}$$

$$T-rd-down \quad \frac{\Gamma_c \vdash x : M \quad \Gamma, y : M_c \vdash P : A}{\Gamma_c \vdash x^z \langle y : M \rangle P : A}$$

$$T-wr-local \quad \frac{\Gamma_c \vdash x : M \quad \Gamma_c \vdash y : M \quad \Gamma_c \vdash P : A}{\Gamma_c \vdash \overline{x}^* \langle y : M \rangle P : A}$$

$$T-wr-up \quad \frac{\Gamma_c \vdash x : M \quad \Gamma_c \vdash y : M \quad \Gamma_c \vdash P : A}{\Gamma_c \vdash \overline{x}^\uparrow \langle y : M \rangle P : (A \oplus [x : M])}$$

$$T-wr-down \quad \frac{\Gamma_c \vdash x : M \quad \Gamma_c \vdash y : M \quad \Gamma_c \vdash P : A}{\Gamma_c \vdash \overline{x}^z \langle y : M \rangle P : A}$$

$$T-declassify \quad \frac{\Gamma, x : M_c \vdash P : A \quad \Gamma_c \vdash u : M' \quad \Gamma_c \vdash Q : A}{\Gamma_c \vdash \text{et } (x : M) = (M' \mapsto M)(u \to u') \text{ in } P \text{ else } Q}$$

$$T-stop \quad \frac{}{\Gamma_c \vdash 0} \qquad T-sub-interface \quad \frac{A \subseteq A'}{A \le A'}$$

$$T-subsumption \quad \frac{\Gamma_c \vdash P : A \quad A \le A'}{\Gamma_c \vdash P : A'}$$

3   Example Analysis

Here we analyze how to make use of secure Seal to ensure the security of mobile computing without reducing the availability of the system through an example. Reference [9], taking the currently widely used VPN applications as an example. VPNs are often used by companies or organizations to transmit confidential information over public networks. According to the requirements of the BLP model, the plaintext sent by the mobile terminal in the VPN is encrypted to form a cipher text to transmit on the common channel. Sender is used to represent the mobile terminal process. Server represents the VPN server process (message receiver) and *msg* is the message. First we analyze the Receiver passes information through the public channel using plaintext, and the communication process with the security Seal calculation is described as follows:

$$Sender \triangleq \overline{pCh}\langle msg : T_{msg}\rangle.0$$
$$Server \triangleq (vsCh : T_{sCh})pCh\langle x : T_x\rangle.\overline{sCh}\langle x : T_x\rangle.0;$$
$$P_{VPN} \triangleq Sender | Server$$

Where pCh is the public channel and sCh is the private channel inside the server. The process *Sender* indicates that the mobile terminal terminates after sending the message *msg* of type T_{msg} on channel pCh; the process *Server* indicates that the server receives the message variable x of type Tx on the channel pCh, And on the channel sCh outputs it to other processes, and then terminated. Limited to space, we do not describe the process of dealing with the message. Thus, a complete VPN communication process PVPN is formed by the process Sender and Server.

Let's take a closer look at the names, variables, and types of processes. According to the characteristics of VPN application and the definition of BLP model, pCh as a public transport channel on the network, the type domain should be {network} with a lower security level, where we set it the public. It is readable and writable, so its type is $\{(\{network\}, public)^r, (\{network\}, public)^w\}$. Msg is of type field Φ, and has a higher security level, where we set it the secret, so the type of msg is $(\Phi, secret)$. The variable x in the process *Server* is mainly used to represent the information *msg* received on the channel pCh, so the type of x can be defined as $(\{network\}, secret)$. Similarly, the type of private channel sCh can be defined as $(\{network, server\}, secret)$. Based on the above definition, we can give the definition of the environment type Γ:

$$\Gamma \triangleq \{msg : T_{msg}, pCh : T_{pCh}\}$$

$$Sender' \triangleq let(x : T_{pCh}) = (T_{msg} \mapsto T_{pCh})(msg \rightarrow msg') \text{ in } \overline{pCh}\langle x : T_{pCh}\rangle.0 \text{ else } 0$$
$$Server \triangleq (vsCh : T_{sCh})pCh\langle x : T_x\rangle.\overline{sCh}\langle x : T_x\rangle.0;$$
$$P'_{VPN} \triangleq Sender' | Server$$

Proposition 1. P_{VPN} is not a good type in Γ.

Proof. Contradiction. Assuming P_{VPN} is a good type, then obviously there are processes Sender and Server in the good type. If the Sender is required to be of a good type, then there must be a type T such that both $\Gamma \vdash msg : T$ and $\Gamma \vdash pCh : T$ are true. From Γ and the definition of subtype relation, it is clear that there is no such type T. Therefore, *Sender* is not a good type, then P_{VPN} is not a good type. Proof finished.

Proposition 1 indicates that the type system in P_{VPN} prevents *Sender* from transferring important messages *msg* directly to the *Server*, which satisfies the requirements of the VPN application and also meets the BLP model requirements in the multi-level security policy. The high security level information cannot be passed to the low security level. To ensure the security of VPN applications, we need to encrypt the high-security level message, that is, by the open operator. This process can be described as a security Seal calculus as follows:

$$Sender' \triangleq \text{let}(x : T_{pCh}) = (T_{msg} \mapsto T_{pCh})(msg \to msg') \text{ in } \overline{pCh}\langle x : T_{pCh}\rangle.0 \text{ else } 0$$

$$Server \triangleq (vsCh : T_{sCh})pCh\langle x : T_x\rangle.\overline{sCh}\langle x : T_x\rangle.0;$$

$$P'_{VPN} \triangleq Sender' | Server$$

Proposition 2. P'_{VPN} is a good type in Γ.

Proof. As the process *Sender'* and *Server* make up the process P'_{VPN}, to prove that P'_{VPN} is good type, you have to prove that the process *Sender'* and *Server* itself is also good type. Obviously, we need to prove: (1) $\Gamma \vdash msg : T_{msg}$ Γ ,x:, (2) $T_{pCh} \vdash \overline{pCh}\langle x : T_{pCh}\rangle.0$, (3) $\Gamma \vdash 0$, according to Γ and the definition of type rules and Specification, it is easy to prove (1) and (3). Here we prove (2) $T_{pCh} \vdash \overline{pCh}\langle x : T_{pCh}\rangle.0$ that this can be proved by.

Proposition 2 indicates that the type system of P'_{VPN} does not prevent *Sender'* from passing the public information *msg'* on the public channel. Compared with the process *Sender*, it is clear that the process *Sender'* transfer the type of message *msg* from private to public. In practical application, the security of *Sender'* is guaranteed by the correctness of $\Gamma \vdash msg : T_{msg}$. Obviously, in VPN applications, we use the encryption algorithm to ensure security, in other words, the type conversion is achieved by the encryption algorithm. Therefore, the security of *Sender'* depends on the security of the encryption algorithm. If the encryption algorithm is safe, obviously *Sender'* is also safe. Therefore, the encrypted information can be transmitted on the public channel. It meets our actual demands for security.

4 Summary

In this paper, a security type system Seal calculus is proposed. Based on the security model of the system, the access control of the distributed system is abstracted into the I/O operation (read and write operation) of the system. The definition of the security policy is transformed into the assignment of the variable type, And with the help of the type system widely used in the programming language, fine-grained access control can be achieve in the programming language level. At the same time, in order to solve the practical problems of channel control and security power reduction, an effective dynamic transformation framework based on mandatory type transformation is proposed. The static detection and dynamic detection are organically integrated to form a unified security model with good scalability and usability. The security type system proposed in this paper is an efficient security indication and implementation method. However, the model only describes the security characteristics of the distributed system, and cannot describe the integrity of the system security features. It will be the focus of our next work.

Acknowledgment. This work is supported by the Key project of science and technology research in Guangxi education (No. 2013ZD021), the innovation team project of of xiangsihu youth scholars of Guangxi University For Nationalities, the Application Research Program of 2016 the Guangxi province of China young and middle-aged teachers basic ability promotion project (No. KY2016YB133), the Research Program of 2014 Guagnxi University for Nationalities of China (No. 2014MDYB029), and the Research Program of 2014 Guagnxi University for Nationalities of China (No. 2014MDYB028).

References

1. Wu, Z.-Z., Chen, X.-Y., Yang, Z., et al.: Survey on information flow control. J. Softw. **28**(1), 135–159 (2017). (in Chinese)
2. Pasquier, T.F.J.M., Singh, J., Bacon, J.: Managing big data with information flow control. SIGARCH Comput. Archit. News **14**(9), 721–731 (2014)
3. Bake, D.B.: Fortresses built upon sand. In: Proceedings of the New Security Paradigms Workshop, pp. 148–153 (1996)
4. Deng, Y., Sangiorgi, D.: Towards an algebraic theory of typed mobile processes. Theor. Comput. Sci. **350**(2–3), 188–212 (2004)
5. Yoon, M.K., Chen, N.S.Y., Christodorescu, M.: PIFT: predictive information-flow tracking. In: Proceedings of the ASPLOS, pp. 246–253. ACM Press, Atlanta (2016)
6. Nardelli, F.Z.: Types for seal calculus. Master thesis (2000)
7. Braghin, C., Sharygina, N., Barone-Adesi, K.: Automated verification of security policies in mobile code. In: Davies, J., Gibbons, J. (eds.) IFM 2007. LNCS, vol. 4591, pp. 37–53. Springer, Heidelberg (2007). https://doi.org/10.1007/978-3-540-73210-5_3
8. Hennessy, M., Riely, J.: Information flow vs. resource access in the asynchronous pi-calculus. ACM Trans. Program. Lang. Syst. **24**(5), 566–590 (2002)
9. Guo, Y.-C., Fang, B.-X., Yin, L.-H., et al.: A security model for confidentiality and integrity in mobile computing. Chin. J. Comput. **36**(7), 1424–1433 (2013). (in Chinese)
10. Li, Q., Yuan, Z.-X.: Permission type system for internal timing information flow in multi-thread programs. Comput. Sci. **41**(3), 163–168 (2014). (in Chinese)

Cloud Computing Resource Scheduling Optimization Based on Chaotic Firefly Algorithm Based on the Tent Mapping

Xiaolan Xie[1,2] and Mengnan Qiu[1(✉)]

[1] College of Information Science and Engineering,
Guilin University of Technology,
Guilin, Guangxi Zhuang Autonomous Region, China
837315716@qq.com
[2] Guangxi Universities Key Laboratory of Embedded Technology
and Intelligent System, Guilin University of Technology, Guilin, China

Abstract. In order to improve the utilization of cloud computing resources and maintain the load balance, this paper proposes a cloud computing resource scheduling optimization chaotic firefly algorithm based on the Tent mapping to solve the problem that the firefly algorithm has premature convergence and is easily trapped in the local optimum. In the firefly algorithm, a chaotic algorithm based on the Tent mapping is introduced. By perturbing individuals, the convergence speed is accelerated and the local most optimal probability is reduced. The Bernoulli shift transformation is introduced to improve the cloud computing model. The simulation results based on CloudSim show that the algorithm can shorten the task completion time and improve the overall processing capability of the system.

Keywords: Firefly algorithm · Tent mapping · Chaos optimization
Cloud computing · Resource scheduling · CloudSim

1 Introduction

The Firefly Algorithm (FA) is developed by simulating the biological characteristics of adult luminescence in nature, and it is also a stochastic optimization algorithm based on group. At present, the firefly algorithm was successfully applied to the optimal solution of the function. In addition, the firefly algorithm also has many uses in other fields. The algorithm has the advantages of simple operation, strong robustness, and easy implementation. However, there are also disadvantages that it is easy to get precocious and fall into local extreme points and shocks. Aiming at these disadvantages, a chaotic firefly algorithm based on Tent mapping is proposed in this paper. The simulation experiments show that the improved firefly algorithm has faster convergence than the original algorithm, can shorten the task completion time and improve the overall processing capability of the system [1].

Since the firefly algorithm was put forward, many scholars at home and abroad have improved the algorithm [2–8]. For example, Yang et al. [9] introduced a chaotic algorithm based on Logistic mapping in the firefly algorithm and introduced the Lagrange slack function to improve the cloud computing model, which can effectively

Q. Chen et al. (Eds.): ATIS 2018, CCIS 950, pp. 118–128, 2018.
https://doi.org/10.1007/978-981-13-2907-4_10

avoid the unbalance of resource allocation and shorten the completion time of the task. Mo et al. [10] used the Simplex Algorithm to improve the success rate of the algorithm in the standard firefly algorithm. Feng et al. [10] proposed a dynamic population firefly algorithm based on chaos theory, which reduced the invalid movement of the firefly and improved the accuracy of the algorithm. Wu et al. [11] improved the biggest attraction in the firefly algorithm and applied it in T-S fuzzy recognition. After these improvements, the firefly algorithm is more suitable for solving many problems such as unconstrained optimization, constrained optimization, multi-objective programming and so on.

In order to make further improvement on the role of firefly algorithm in cloud computing resources scheduling, this paper designs an improved firefly algorithm to apply chaos on individuals in the worst position of firefly individuals. This paper uses Tent mapping as a chaos optimization model to optimize. Finally, simulation experiments verify the effectiveness of this improved method.

2 Establishment of Resource Scheduling Model in the Cloud Computing

In the cloud computing environment, tasks and computing resources do not have a direct one-to-one relationship. Instead, tasks are first mapped to resources, and then resources are mapped to corresponding physical devices [12–15]. Most current cloud computing environments adopt the MapReduce programming model proposed by Google. The implementation process is shown in Fig. 1 [16–18].

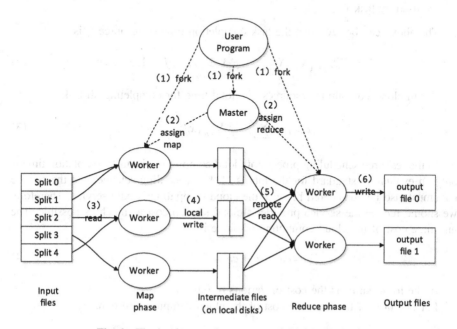

Fig. 1. The implementation process of the MapReduce

Before setting up a resource scheduling model on cloud computing, make the following assumptions:

(1) The performance of a virtual machine (resource) can achieve the requirements of any task.
(2) All tasks can be fully allocated.
(3) A task can only be assigned to one resource.

In the cloud computing system, suppose there were m resources and n users in all, then the model of resource scheduling on cloud computing can be described as

$$M = \{U, R, F, A\} \tag{1}$$

In the formula: U is a set of users. R represents a set of resources. F is an objective function. A is a solution algorithm. This paper selects IQPSO algorithm.

The concrete features of the resource scheduling model on cloud computing are as follows [16]:

(1) According to core count, memory size and disk space, the i-th resource (r_i) can be subdivided, and it can be expressed as $r_i = \{\lambda_i, \mu_i, \varphi_i\}$. Then resources can be expressed as $R = \{r_1, r_2, \ldots, r_m\}$.
(2) A user has n tasks. Each task is same and mutually independent. Then all the task sets can be described as: $T = \{t_1, t_2, \ldots, t_n\}$. The time of the task execution is $S_{m \times n} = \{s_{ij}\}$. In it, s_{ij} represents the execution time of the task i in resource j.
(3) The source scheduling matrix E: $E_i = (e_i)$ represents that resource e_i performs task i. The source using matrix $X_{m \times n} = \{X_{je_i}\}$. In it, X_{je_i} represents that resource e_i is used by task j.

The above can be seen that the task completion time of resource r_i is

$$T_j = \max_{1 \leq i \leq n} \{X_{ij} \cdot X_{je_i}\} \quad i = 1, 2, \ldots, n, j = 1, 2, \ldots, m \tag{2}$$

For m cloud computing resources, the total time for completing all tasks is

$$T_{total} = \sum_{j=1}^{m} \max_{1 \leq i \leq n} \{x_{ij}, x_{je_i}\} \tag{3}$$

In the resource scheduling process of cloud computing, the task completion time is only an evaluation standard for the merits of the scheduling scheme, and the service cost must also be considered because the cloud computing is paid service. Therefore, we should reduce the service provider's cloud service cost as much as possible. The unit time cost of the cloud computing resource is

$$C_j = c_{i1} \times \alpha_i + c_{i2} \times \beta_i + c_{i3} \times \gamma_i \tag{4}$$

In the formula: c_{ij} is the cost of various resources.
For one task of the user, the cost of the cloud computing resource r_i is

$$C_j = (c_{i1} \times \alpha_i + c_{i2} \times \beta_i + c_{i3} \times \gamma_i) \times \sum_{j=1}^{m} \max_{1 \leq i \leq n} \{x_{ij}, x_{je_i}\}$$
$$= C_j \times \max_{1 \leq i \leq n} \{x_{ij}, x_{je_i}\} \tag{5}$$

For all tasks of the user, the total cost that cloud service provider executes is

$$C = \sum_{j=1}^{m} C_j \times \max_{1 \leq i \leq n} \{x_{ij}, x_{je_i}\} \tag{6}$$

In this way, the cloud computing resource scheduling and the optimized objective function is to minimize the executing total cost of the cloud service provider. That is min(C).

From Fig. 2, we can see that MapReduce implementation mainly includes these steps:

(1) Slice the input data source first.
(2) Master schedules worker and executes the map task.
(3) Worker reads the input source clip.
(4) Worker executes the map task and save the output task locally.
(5) Master schedules worker and executes the reduce task, reduce worker reads the output files of the map task.
(6) Execute the reduce task and save the output task into the HDFS.

3 Resource Scheduling Algorithm in the Cloud Computing

3.1 Improved Chaotic Firefly Algorithm

The firefly algorithm is developed to simulate the biological characteristics of adult luminescence in nature, and it is also a population-based stochastic optimization algorithm. The Firefly algorithm is a heuristic algorithm inspired by the behavior of fireflies at night. After several years of the development, the firefly algorithm has a good application prospect in the optimization process of the continuous space and some production scheduling.

The mathematical description of the algorithm is as follows:

(1) Initialization

In the feasible field, n fireflies are randomly placed, and each firefly's fluorescein is l_0. Dynamic decision domain is r_0. Initialization step is s. Domain threshold is n_t. The disappearance rate of the fluorescein is ρ. The update rate of the fluorescein is γ. The update rate of the dynamic decision domain is β. The perceptual domain of fireflies is r_s. The number of the iteration time is M.

(2) Update the fluorescein of the firefly i

$$l_i(t) = (1 - \rho)l_i(t - 1) + \gamma J(x_i(t)) \tag{7}$$

In it, $J(x_i(t))$ denotes the value of the objective function of the firefly i at time t. $l_i(t)$ denotes the value of the fluorescein of the firefly i at time t.

(3) Find the neighbors of the firefly i

$$N_i(t) = \{j : ||x_j(t) - x_i(t)|| < r_d^i(t); l_i(t) < l_j(t)\} \tag{8}$$

In it, $N_i(t)$ denotes the collection of neighbors of the firefly i at time t. $r_d^i(t)$ denotes the dynamic decision domain of the firefly i at time t.

(4) Determine the direction of movement of the firefly i

$$j = \max(p_i)$$

In it, $p_i = (p_{i1}, p_{i2}, \ldots, p_{iN_i(t)})$.
The probability of the shift is

$$p_{ij}(t) = \frac{l_j(t) - l_i(t)}{\sum_{k \in N_i(t)} (l_k(t) - l_i(t))} \tag{9}$$

(5) Update the location of the firefly i

$$X_j(t+1) = X_j(t) + s\left(\frac{X_j(t) - X_j(t)}{||X_j(t) - X_j(t)||}\right) \tag{10}$$

(6) Update the dynamic decision domain

$$r_d^i(t+1) = \min\{r_s, \max\{0, r_d^i(t), \beta(n_t - |N_i(t)|)\}\} \tag{11}$$

Chaos has the ergodicity and intrinsic randomness of the phase space. It uses chaotic variables to perform optimal search, so that it can jump out of local optimum and achieve global optimization. With the increase of the number of the iteration time, the firefly individuals are getting closer to $X_j(t)$, which results in the loss of differences between individuals. In order to prevent the occurrence of this phenomenon, chaos is performed on individuals in the worst position of firefly individuals. The Tent mapping is used as the chaos optimization model in this paper. The Tent mapping have an expression that

$$g(x) = \begin{cases} 2x, & 0 \le x \le \frac{1}{2} \\ 2(1-x), & \frac{1}{2} < x \le 1 \end{cases}$$

The Tent mapping can be expressed by a Bernoulli shift.

$$g(x) = \begin{cases} 2x, & 0 \le x \le \frac{1}{2} \\ 2x - 1, & \frac{1}{2} < x \le 1 \end{cases}$$

That is

$$x_{n+1} = g(x_n) = (2x_n)mod1 \tag{12}$$

For the floating point between zero and one, when the computer performs the Tent mapping, it is actually moving the binary number of the decimal part to the left. This kind of computing feature makes full use of the characteristics of the computer and is more suitable for the processing of data sequences in the big data level. Therefore, the Tent mapping has faster iteration than the Logistic mapping.

The strategy process of the chaotic optimization firefly algorithm is as follows [9]:

Step1 Suppose $x_i = (x_{i1}, x_{i2}, \ldots, x_{in})$. x_i is mapped to the value of optimization variable in the firefly algorithm. In it, x_{min} and x_{max} represent the minimum and maximum value of the variables respectively.

$$y_i = x_{min} + (x_{max} - x_{min}) \times x_i \tag{13}$$

Step2 The formula (12) is done the multiple iterations of chaotic sequence sets, which is y_i^m.

Step3 According to the principle of the pre-image, introducing the formula(13). The set of feasible solutions obtained is $x_i^{(m)}$.

$$x_i^{(m)} = x_{min} + (x_{max} - x_{min}) \times y_i^m \tag{14}$$

Step4 After the chaotic mapping is adopted, the firefly individual passes some individuals of the probability q. In it, t is the current number of the iteration. The formula of q is

$$q = \frac{\ln t}{(1 + \ln t)} \tag{15}$$

3.2 The Description of the Scheduling Algorithm on the Resource Optimization

The steps of the solution are as follows.

(1) The initial number of fireflies group is N. The maximum number of the iteration is max.

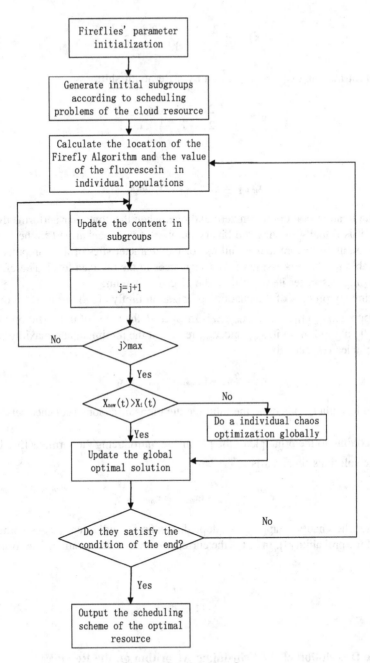

Fig. 2. The algorithm flow

(2) Individuals in the firefly algorithm are coded, and the fluorescein value of the individual firefly is calculated according to the Eqs. (10) and (7). In order to prioritize the results, the firefly population is divided into m subgroups.

(3) Update the value of the fluorescein in each subgroup. Specifically,

① According to the formula (10), finding their best position and worst position.

② Calculating the location of the firefly according to the formula (7). Then updating the firefly according to the formula between twelve and fifteen, and getting the new position of the new individual $(x_{new}(t), x_i(t))$.

③ Calculating $x_{new}(t)$ of each firefly. If $x_{new}(t) > x_i(t)$, else $i = new$, otherwise updating again.

④ If the number of the iteration is less than max, going to the step (1).

(5) If the termination condition is satisfied, the optimization process ends, otherwise, it proceeds to (2) to continue optimization.

(6) According to the most optimal individual, the most optimal cloud computing scheduling scheme is obtained.

The algorithm flow is shown in Fig. 2.

4   Simulation Experiment

The simulation platform used in this paper is Core CPU 3.0 GHz, 2 GB DDR 3 and Windows XP operating system. CloudSim simulation software is used for the simulation experiment. In order to verify the feasibility and performance of the improved algorithm proposed in this paper, we compare it with the basic firefly algorithm under the same conditions.

Under the simulation environment set up by the CloudSim, ten scheduling task is taken as an example to test the time of the task completion. Table 1 shows the total time required to complete 10 tasks with the sequential assignment strategy.

The total time required to complete 10 tasks with the improved firefly algorithm is shown in Table 2.

Table 1. The total time required to complete 10 tasks with the sequential allocation strategy

Task number	Execution status	Data center number	Virtual machine number	Execution time/s	Starting time/s	Completion time/s
0	Success	1	0	67.54	0	67.54
1	Success	1	1	171.83	0	171.83
2	Success	1	2	226.91	0	226.91
3	Success	1	3	209.36	0	209.36
4	Success	1	4	56.79	0	56.79
5	Success	1	0	64.26	67.54	131.80
6	Success	1	1	68.88	171.83	240.71
7	Success	1	2	236.95	226.91	463.86
8	Success	1	3	145.93	209.36	355.29
9	Success	1	4	106.31	56.79	163.10

Table 2. The total time required to complete 10 tasks with the improved firefly algorithm

Task number	Execution status	Data center number	Virtual machine number	Execution time/s	Starting time/s	Completion time/s
0	Success	1	0	66.79	0	66.79
1	Success	1	1	169.56	0	169.56
2	Success	1	2	225.21	0	225.21
3	Success	1	3	207.85	0	207.85
4	Success	1	4	55.12	0	55.12
5	Success	1	0	62.74	66.79	129.53
6	Success	1	1	67.34	169.56	236.90
7	Success	1	2	233.58	225.21	458.79
8	Success	1	3	142.36	207.85	350.21
9	Success	1	4	104.53	55.12	159.65

It can be seen from the Table 1 and the Table 2 that the task scheduling with the improved firefly algorithm greatly shortens the time of the task completion.

In addition, after comparing the improved firefly algorithm with the sequential algorithm, the greedy algorithm and basic firefly algorithm in dealing with large-scale operation problems, we find that the improved firefly algorithm is greatly shortened the time of the task compared to other algorithms with the increase of the number of tasks. It is shown in Fig. 3.

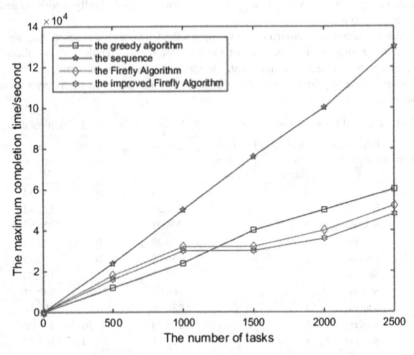

Fig. 3. The time for completing large-scale tasks under four different task scheduling strategies

It can also be seen from Fig. 3 that the improved firefly algorithm has better convergence than other algorithms.

5 Conclusion

In the cloud computing environment, how to allocate and use resources more reasonably has always been a research hotspot. For the problem that the basic firefly algorithm solves the problems with slow speed and is easy to fall into the local optimum, this paper introduces a chaotic algorithm based on the Tent mapping in the firefly algorithm, and it aims at resource scheduling model of the cloud computing, which uses the Bernoulli shift transformation for simplification. Simulation results show that the proposed algorithm can effectively schedule cloud computing resources and solve the problem of uneven resource allocation in the cloud computing, and it has good convergence performance, which makes the task parameters to achieve optimal.

Acknowledgements. This research work was supported by the National Natural Science Foundation of China (Grant No. 61762031), Guangxi Key Research and Development Plan (No. 2017AB51024, 2018AB8126006), GuangXi key Laboratory Fund of Embedded Technology and Intelligent System.

References

1. Zuo, Z., Guo, X., Li, W.: An improved swarm optimization algorithm. Microelectron. Comput. **35**(2), 61–66 (2018)
2. Jia, Y., Liu, J.: Optimization and application of firefly algorithm based on CloudSim. J. Beijing Inf. Sci. Technol. Univ. **33**(1), 66–70 (2018)
3. Li, L., Yao, Y., Li, T.: Study on improved artificial firefly algorithm in cloud computing resources. Appl. Res. Comput. **30**(8), 2298–2333 (2013)
4. Li, J., Peng, J.: Task scheduling algorithm based on improved genetic algorithm in cloud computing environment. J. Comput. Appl. **31**(1), 184–186 (2011)
5. Wang, F., Li, M., Daun, W.: Cloud computing task scheduling based on dynamically adaptive ant colony algorithm. J. Comput. Appl. **33**(11), 3160–3162 (2013)
6. Ye, S., Wenbo, Z., Hua, Z.: SLA-oriented virtual resources scheduling in cloud computing environment. Comput. Appl. Softw. **32**(4), 11–17 (2015)
7. Sun, H., Zhu, J.: Design of task-resource allocation model based on Q-learning and double orientation ACO algorithm for cloud computing. Comput. Meas. Control **22**(10), 3343–3347 (2014)
8. Shen, J., Wu, C., Hao, Y., Yin, B., Lin, Y.: Elastic resource adjustment method for cloud computing data center. J. Nanjing Univ. Sci. Technol. **39**(1), 89–93 (2015)
9. Yang, D., Li, C., Yang, J.: Cloud computing resource scheduling based on improving chaos firefly algorithm. Comput. Eng. **41**(2), 17–20 (2015)
10. Mo, Y., Ma, Y., Zheng, Q., et al.: Improved firefly algorithm based on simplex method and its application in solving non-linear equation groups. CAAI Trans. Intell. Syst. **9**(6), 747–755 (2014)
11. Wu, D., Ding, X.: T-S model identification based on improved firefly algorithm. Comput. Simul. **30**(3), 327–330 (2013)

12. Zhang, H., Chen, P., Xiong, J.: Task scheduling algorithm based on simulated annealing ant colony algorithm in cloud computing environment. J. Guangdong Univ. Technol. **31**(3), 77–82 (2014)
13. Lan, F., Yong, Z., Ioan, R., et al.: Cloud computing and grid computing 360-degree compared. In: Proceedings of Grid Computing Environments Workshop, pp. 268–275. IEEE Press (2008)
14. Sesum-Cavic, V., Kuhn, E.: Applying swarm intelligence algorithm for dynamic load balancing to a cloud based call center. In: Proceedings of the 4th IEEE International Conference on Self Adaptive and Self Organizing Systems, pp. 255–256. IEEE Press (2010)
15. Grossman, R.L.: The case for cloud computing. IT Prof. **11**(2), 23–27 (2009)
16. Zhao, L.: Cloud computing resource scheduling based on improved quantum partical swarm optimization algorithm. J. Nanjing Univ. Sci. Technol. **40**(2), 223–228 (2016)
17. Dean, J., Ghemawat, S.: Map/reduce: simplified data processing on large clusters. Commun. ACM **51**(1), 107–112 (2008)
18. Zhang, H., Han, J., Wei, B., Wang, J.: Research on cloud resource scheduling method based on map-reduce. Comput. Sci. **42**(8), 118–123 (2015)

An Optimized Parallel Algorithm for Exact String Matching Using Message Passing Interface

Sucharitha Shetty[1] , B. Dinesh Rao[2(✉)] , and Srikanth Prabhu[1]

[1] Department of Computer Science and Engineering,
Manipal Institute of Technology, Manipal Academy of Higher Education,
Manipal 576104, Karnataka, India
sucha.shetty@manipal.edu
[2] School of Information Science, Manipal Academy of Higher Education,
Manipal 576104, Karnataka, India
dinesh.rao@manipal.edu

Abstract. Data Partitioning is one of the key approaches followed in parallelism. It has proved its efficiency on many algorithms and still lot of work is going on this area. In this paper we propose an optimization technique for string matching using data partitioning with multi-core architecture. The paper primarily focuses on caching and re-utilization of processes. The experiments showed that concept of caching increased the speed drastically for frequently asked patterns. The MPI proposed implementation highlighted the increase in efficiency using multi-core and decrease in performance when the cores were reutilized.

Keywords: Data partitioning · MPI I/O

1 Introduction

Huge amount of information needs to be processed every moment which requires enhancing the performance of the systems. This can be achieved either by increasing the clock frequency or increasing the number of operations to be performed in each clock cycle. High Performance Computing System addresses this requirement by performing voluminous operations per clock cycle. The main feature of this system is parallelism. Techniques like using multiple ALU's in a single CPU, a single memory with multiple processors for increasing the execution of a number of instructions in a unit time or cluster of computers acting as a single processing unit that can share the load on the program are adopted. A number of applications have been devised using parallelism.

String matching is one such researched topic because of its wide applications in web search engines, library systems, network intrusion detection system, speech and pattern recognition, DNA sequencing and various other fields [9].

Startup, interference, and skew [1] are the barriers for speed enhancement of parallel computation. Startup is the time needed to start a parallel operation. Interference

Q. Chen et al. (Eds.): ATIS 2018, CCIS 950, pp. 129–135, 2018.
https://doi.org/10.1007/978-981-13-2907-4_11

happens when a process is slow compared to others while using a shared resource. Skew occurs as the number of parallel step increases the serviced time of the final job decreases.

In this paper, we present an optimized parallel algorithm for solving the string matching problem. This algorithm makes use of shared file pointers using MPI I/O routines which follows the SPMD (Single Program Multiple Data) model. The work also analyses the mentioned barriers under various scenarios.

The rest of the paper is organized as follows. Related work, background on the MPI I/O and data parallelism is addressed in Sect. 2. Section 3 presents our proposed approach. Experimental results and the performance analysis is presented in Sect. 4. The conclusion is described in Sect. 5.

2   Survey

2.1   Related Work

Navarro [5] has addressed the various types of search patterns like searching fixed-length string, optional characters, strings within some length range and usage of regular expressions. The work considers exact as well as approximate pattern matching using sequential searching and indexed searching.

Rajashekharaiah et al. [6] have implemented exact string matching using grid MPI parallel programming model. The data was partitioned among tasks and then using MPI collective communication the data was consolidated. It was observed that the speedup and execution time increased compared to sequential.

Sharma et al. [7] have implemented a parallel linear search algorithm on numbers by using four values of the array, two searches from the beginning of the array and two from the back of the array. It is observed that this parallel approach is faster than linear search.

Byna et al. [8] have explored the performance benefits of parallel I/O prefetching. The work concentrates on fetching data quickly based on frequently requested patterns and thereby addressing future I/O demands. These are stored in prefetch cache.

2.2   MPI I/O

MPI [3] has introduced the idea of parallel I/O in version two of MPI specification. However, its use by end user is less, but research has proved that using shared file systems, MPI I/O can significantly improve the I/O performance.

The MPI I/O data access routines differ in a lot of aspects to the regular routines of MPI [2]. One among them is the identification of offset of each process to access the shared file: using file pointers or by explicitly calculating the offset. Another aspect is the synchronization between the independent operations and collective operations. One of the reasons for the less usage is that it has been observed that [4] all known methods provide a narrow support for shared file pointers, depending on the underlying file system and the platform.

2.3   Data Parallelism

Flynn's classification has SIMD (Single Instruction Multiple Data) [10] as one of its classes as shown in Fig. 1. It is simpler compared to the other three classes as it follows the data parallelism approach. SIMD executes a single instruction at any point of time on all elements in the data set. Vector processors and GPU make use of this concept.

	Single instruction	Multiple instructions
Single data	SISD	MISD
Multiple data	SIMD	MIMD

Fig. 1. Flynn's taxonomy

3   Methodology

This section discusses the proposed algorithm which has modified the earlier parallel linear search algorithm [9] using MPI. The master-slave algorithm had a master which broadcasted the pattern and text to workers. The workers saved the sub-texts in their memory and perform the sequential string matching operation. All the workers send their count of matches to the master who consolidates and gives the final count.

Compared to the above algorithm, the proposed work assumes exact string matching on an unmodified file with sorted strings using linear search algorithm. The work is a comparison of three optimizations made to the above-mentioned algorithm. Here, the master broadcasts the pattern and name of the shared file. An offset for each of the processors is calculated. Every processor now has a region of text to search for the pattern. However, since the file is sorted in the proposed algorithm, some processors may have the area of the search pattern while others may not hence they need not process the data. This is identified based on the first letter of the first word and the first letter of the last word of the allotted area of the file to each processor, only those processors which satisfied the range within which the first letter of the pattern could appear were the important areas. These were the processors which searched for the pattern whereas the other processors were not utilized. This was one approach as shown in Figs. 2 and 3 where the search operation was speeded up.

Another improvement was the concept of caching. If the search pattern was queried earlier as shown in Fig. 2, then its data is made available in the cache. The output is displayed from the cache. This saved a lot of time for re-searching the pattern with the assumption that the file is not modified.

The third approach was instead of making other processors idle as implemented in the first approach, they are also involved by further partitioning the data of the processor which was identified for having the pattern and distributed among the idle processors as shown in Fig. 4. A linear string matching procedure then operated on every processor. Each processor completes its string matching operation on its local area and returned the number of occurrences. The master then calculated the number of occurrences.

The proposed flow chart for the master-slave model is as follows:

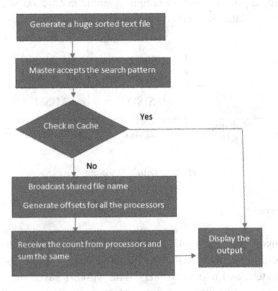

Fig. 2. Flowchart of the master with caching

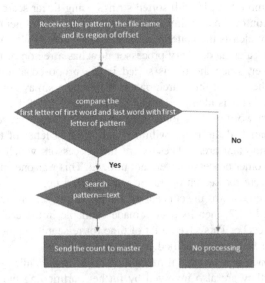

Fig. 3. Flowchart of the first approach with identified processor processing

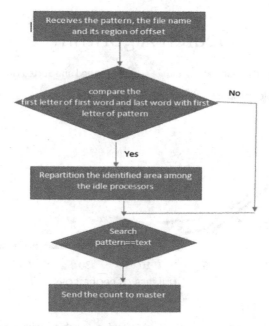

Fig. 4. Flowchart of the second approach with further re-partitioning

4 Experimental Results and Analysis

In this section, we present the experimental results for the performance of the parallel string matching implementation using our proposed approaches. The target platform for our experimental study is a high performance computer with two INTEL Xeon Phi 7120A accelerator cards each with 60 cores.

The experiment was conducted on sequential linear search algorithm as well. It was observed that there is a similarity between sequential and parallel algorithm with one processor since though good parallel algorithm was written the partitioning code could not function that efficiently since it made use of single processor. However, processing on sorted data was faster.

Reduction in speed was achieved by using the concept of cache, if the file is an unmodified one, the results of frequently accessed patterns are stored in RAM. This helped in the fast retrieval of data. The proposed algorithm of only processing those processors which may contain the pattern achieved good speedup. But, when further reuse of partition was applied, there was not much improvement in speed. The conclusion made with the third approach was parallel concept works fine for single program on multiple data but as the number of parallel steps increased with further re-utilizing the processors, the performance degraded. These three approaches were plotted and the results are shown in Fig. 5.

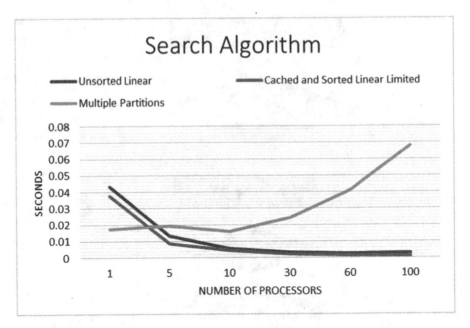

Fig. 5. Experimental execution time (in secs) of string matching with respect to number of processors for text size of 12 MB using several pattern lengths

5   Conclusion

In this paper, an optimized approach for parallel linear search is proposed. The experiments have shown an increase in efficiency and speed improvement with the use of caching. The work has demonstrated that adding further logic to the parallel and reutilization of the processor reduces the speed.

References

1. DeWitt, D., Gray, J.: Parallel database systems: the future of high performance database systems. Commun. ACM **35**(6), 85–98 (1992)
2. Chaarawi, M., Dinan, J., Kimpe, D.: On the usability of the MPI shared file pointer routines. In: Träff, J.L., Benkner, S., Dongarra, J.J. (eds.) EuroMPI 2012. LNCS, vol. 7490, pp. 258–267. Springer, Heidelberg (2012). https://doi.org/10.1007/978-3-642-33518-1_31
3. Message Passing Interface Forum: MPI-2: Extensions to the Message Passing Interface (1997). http://www.mpi-forum.org/
4. Chaarawi, M., Gabriel, E., Keller, R., Graham, R.L., Bosilca, G., Dongarra, J.J.: OMPIO: a modular software architecture for MPI I/O. In: Cotronis, Y., Danalis, A., Nikolopoulos, D.S., Dongarra, J. (eds.) EuroMPI 2011. LNCS, vol. 6960, pp. 81–89. Springer, Heidelberg (2011). https://doi.org/10.1007/978-3-642-24449-0_11
5. Navarro, G.: Pattern matching. J. Appl. Stat. **31**(8), 925–949 (2004)
6. Rajashekharaiah, K.M.M., MadhuBabu, C., Raju, S.V.: Parallel string matching algorithm using grid. Int. J. Distrib. Parallel Syst. **3**(3), 21 (2012)

7. Sharma, R., Kumar, R.: Design and analysis of parallel linear search algorithm. Int. J. Latest Trends Eng. Technol. **10**(1), 35–38 (2018)
8. Byna, S., Chen, Y., Sun, X.H., Thakur, R., Gropp, W.: Parallel I/O prefetching using MPI file caching and I/O signatures. In: Proceedings of the 2008 ACM/IEEE Conference on Supercomputing, p. 44. IEEE Press (2008)
9. Michailidis, P.D., Margaritis, K.G.: String matching problem on a cluster of personal computers: experimental results. In: Proceedings of the 15th International Conference Systems for Automation of Engineering and Research, pp. 71–75 (2001)
10. Flynn, M.J.: Very high-speed computing systems. Proc. IEEE **54**(12), 1901–1909 (1966)

Research on Algorithms for Planted (l,d) Motif Search

Li Anya[⊠]

School of Computer Science, Xi'an Shiyou University, Xi'an, China
303773892@qq.com

Abstract. As one of the most challenging problems in bioinformatics, motif search has important implications for gene discovery and understanding of gene regulatory relationships. Planted (l,d) motif searching (PMS) is a widely accepted issue model in the field of motif search. Solving PMS issues involve computer science, bioinformatics and other related knowledge. Of course, a huge amount of computation is probably unavoidable. Designing an effective and optimized method is particularly important for solving the puzzle.

This article mainly introduces and implements related algorithms for planted (l,d) motif search. Firstly, according to the different implementation methods, it can be divided into two categories: enumeration algorithm and local search algorithm. Then four algorithms are proposed to solve the issue, pseudocodes are also available. Finally, in the experimental part, PMSP algorithm has the best performance through programing implementation and operating results evaluation. In the following work, by combining the idea of MapReduce parallel computing, the design of PMSPMR algorithm can achieve more excellent operational results.

Keywords: Planted (l,d) motif search · Execution time · MapReduce

1 Introduction

Motif searching is a basic issue in bioinformatics, it is of great significance for discovering DNA regulatory signals and interpreting the regulatory codes in the genome. Motif refers to a conserved pattern in DNA sequence, which is usually derived from transcription factor binding sites and can be recognized by transcription factors. Generally, there will be many instances of motifs in the regulatory regions of the DNA sequence, which are new sequences resulting from mutations in certain nucleotide sites. In simple terms, if the length of the motif and its instance is l, the maximum number of degenerate positions that the motif appears in the sequence is **d**, each input sequence contains a motif instance, so motif search can be formally defined as planted (l,d) motif search issue.

The PMS issue is a widely accepted problem model for motif search, given the set $D = \{s_1, s_2, \ldots, s_t\}$ of **t** DNA sequences with length **n** on an alphabet, integers l and **d** satisfying $0 \leq d < l < n$. The goal is to find a l-mer (a string of length l) **m** in set D, each sequence s_i contains a l-mer m_i which has at most **d** position differences compared

© Springer Nature Singapore Pte Ltd. 2018
Q. Chen et al. (Eds.): ATIS 2018, CCIS 950, pp. 136–143, 2018.
https://doi.org/10.1007/978-981-13-2907-4_12

to **m**. l-mer **m** is called a (l,d) motif, l-mer m_i in the sequence is called motif search or motif instance of **m**.

The specific work of this thesis can be divided into 4 parts: In the first section there is a general introduction, in the second section, current motif search algorithms are summarized as the idea of enumeration and local search. In the third section, several basic algorithms for motif search issues are described in detail and pseudocodes are added to explain them. In the fourth section, the motif search algorithms previously mentioned are compared in terms of execution time and spatial performance, optimization scheme is proposed based on MapReduce parallel computation.

2   Algorithm Overview

The concept of motif was first proposed by Doolittle RF in 1981, in order to get more unknown human secrets from genetic codes, many searching algorithms for motifs were developed. According to the different ways of implementation, motif search algorithms can be divided into two categories: enumeration algorithms and local search algorithms.

The purpose of enumeration algorithms is to determine motif instance in each sequence, basic idea is to exhaustively traverse the entire search space corresponding to all candidate motifs, evaluate the degree of conservation of each candidate motif in turn, and then output the highest score motif. The computation of such algorithms needs to cover a countable set consisting of instances of motifs and calculation amount is very considerable. For example, in the motif search issue of length l, a set of 4^l l-mers needs to be calculated. For specific motif search, the output of different enumeration algorithms are identical, so time performance is the most important index to assess enumeration algorithm.

Local search algorithms use heuristic search, clustering, probabilistic analysis and statistical methods to construct a plurality of initial models of the motif, then iteratively updates each model to obtain a local optimal solution. It overcomes the disadvantage of high computational complexity brought about by the method of enumeration thinking. It generally solves the problem in a short time, but it does not always guarantee the global optimal solution. Local search algorithms can complete the computation within acceptable time, recognition accuracy is the most important index to assess them.

3   PMS Related Algorithm

3.1   Depth-First Search Based on Candidate Motif Instance Strings

The PMS algorithm based on depth-first search of candidate motif instance strings (Algorithm 1) operates on a string of length l in a depth-first manner and finally outputs a motif and motif instance. At first, a tree with branch nodes $n - l + 1$ is created in a data set with a sequence length of **n**, the depth of the tree is **t**, then use depth-first search to initialize the established tree.

In the depth-first search, it is necessary to judge the string of length l in next sequence to determine whether it meets Hamming distance ≤ **2d** with the string existing in front tree. If satisfied, these strings with a length of l may be motif instances belonging to the same motif. If not satisfied, they are absolutely impossible to belong to the same motif instance. By the way, a pruning strategy is needed to reduce the time and space complexity of the program, directly subtract unnecessary strings and stop traversing. When the traversal depth of the program is **t**, it indicates that **t** motif instances have been found and they meet Hamming distance between each other **2d**, candidate motifs can be determined by the method of site comparison, after judging the **t** l-mers, if they conform the definition of the motif search, motif and motif instance will be output.

The following is a pseudocode description of Algorithm 1:

Algorithm 1: The PMS algorithm based on depth-first search of candidate motif instance strings

---

Input: t, n, l, d
Output: a set of potential motifs Q
1: set a stack stk
2: **while** stack.size() != empty **do**
3:    vector<string> $temp$ ← stk.pop()
4:    **if** temp.size() $< t$ **then**
5:        **if** each i $(0 \leq i \leq n-l+1)$,exists $x_i \in s_i$ such that $d_H(x_i, temp) \leq 2d$ **then**
6:            push x_i to temp
7:    **else**
8:        **if** exists m satisfying motif($temp$) **then**
9:            Add m to Q
10: return Q

3.2   Depth-First Search Based on Candidate Motif Character

This algorithm is similar to Algorithm 1, which operates on the input sequence in the form of a stack, although their essence is different. Algorithm 1 mainly operates on **t** motif instances and finds candidate motifs that meet the conditions, the algorithm in this section first traverses all motifs added with the pruning algorithm and directly determines the qualified motifs.

The algorithm uses stack to search motif, it creates a tree with a depth of l and a number of branches of **4**. Same as Algorithm 1, depth-first search is used to traverse the motif, however, unlike Algorithm 1, the PMS algorithm based on depth-first search of candidate motif character (Algorithm 2) directly traverses the motif instead of traversing the string in motif instance. After completing the traversal, all possible motif collections can be obtained without any omission. While traversing the motif, special attention should be paid to the use of pruning method to remove unqualified motifs. A candidate modal body with length ≤ l is compared with the initial sequence, if the candidate motif and the same length of the initial sequence Hamming distance ≤ **d** cannot be satisfied, it shows that the candidate motifs do not meet the requirements, prune them and finally export all eligible candidate motifs.

The following is a pseudocode description of Algorithm 2:

Algorithm 2： The PMS algorithm based on depth-first search of candidate motif character

---

Input: t, n, l, d
Output: a set of potential motifs Q
1: set a stack stk
2: push A,C,G,T to stk
3: **while** stack.size() != empty **do**
4:    string $temp \leftarrow stk$.pop()
5:    **if** temp.size() $< l$ **then**
6:       **if** exists m $\in_l$ temp satisfying d_H(m_initialSeq, m) $\leq d$ **then**
7:          $temp$ + A/C/G/T, and push to stk
8:       **else**
9:          **if** each i $(0 \leq i < t)$ exists $z_i \in_l$ m_initialSeq[i] satisfying d_H ($temp, z_i$) $\leq d$ **then**
10:             Add $temp$ to Q
11: return Q

3.3 Breadth-First Search Based on Candidate Motif Character

The PMS algorithm based on breadth-first search of candidate motif character (Algorithm 3) is the same as the basic idea of Algorithm 2. The difference is that one is to perform depth-first search for traversal, while the other is to use breadth-first search to traverse. Remarkably, in the process of traversal, a queue is defined to operate on candidate motif when a candidate motif meets the condition of Hamming distance $\leq$ **d**, the program will operate it in the form of a queue.

The following is a pseudocode description of Algorithm 3:

Algorithm 3： The PMS algorithm based on breadth-first search of candidate motif character

---

Input: t, n, l, d
Output: a set of potential motifs Q
1: set a queue que
2: add A,C,G,T to que
3: **while** que.size() != empty **do**
4: string $temp \leftarrow que$.front()
5:    **if** temp.size() $< l$ **then**
6:       **if** exists m $\in_l$ temp satisfying d_H(m_initialSeq, m) $\leq d$ **then**
7:          $temp$ + A/C/G/T, and add to stk
8:    **else**
9:       **if** each i $(0 \leq i < t)$ exists $z_i \in_l$ m_initialSeq[i] satisfying d_H ($temp, z_i$) $\leq d$ **then**
10:          Add $temp$ to Q
11: return Q

3.4 PMSP Algorithm

The PMSP algorithm differs greatly from the previous three algorithms. Its idea is as follows: generate a motif collection for each string **x** of length **l** in s_1, the Hamming distance of l-mer **x** between all motifs in the collection is less than **d**. It is called a candidate motif collection. Each candidate motif **y** in the candidate collection is compared with each l-mer **x'** in s_2-s_t, determine whether there is an l-mer in s_i that has a Hamming distance from **y** which is not greater than **d**. If this l-mer exists in every sequence in s_2-s_t, it indicates that the candidate motif **y** is a potential motif and l-mer **x'** is a potential motif instance, otherwise, select the next l-mer from s_1, regenerate its candidate motif collection and perform the above judgment process until all l-mers in s_1 are traversed.

The following is a pseudocode description of PMSP algorithm:

PMSP Algorithm

Input: t, n, l, d
Output: a set of potential motifs Q
1: $M \leftarrow \Phi // \Phi$ is an empty set
2: **for** each $x, x \in_l s_1$ **do**
3:    **for** each $s_i (2 \leq i \leq t)$ **do**
4:      $C(x, s_i) \leftarrow \Phi$
5:      **if** for $y \in_l s_i$, satisfying $d_H(x, y) \leq 2d$ **then**
6:        Add y to $C(x, s_i)$
7:      **for** each $x' \in_l B_d(x)$ **do**
8:        **if** for each $i (2 \leq i \leq t)$, exists $z_i \in_l C(x, s_i)$ such that $d_H(x', z_i) = d$ **then**
9:          Add x' to Q
10: return Q

---

4   Experimental Evaluation

In this section, above mentioned four algorithms are code-implemented respectively and the execution time of algorithms is recorded to compare. During the experiment, the values of (l,d) were (10,1), (11,2), (12,3), (14,4) respectively and the execution time of programs were counted. The following is a record table for the execution time of programs (Table 1).

Table 1. Comparison of four PMS algorithms.

(l,d)	Algorithm 1	Algorithm 2	Algorithm 3	PMSP
(10,1)	12 s	161 s	160 s	19 s
(11,2)	61 s	2753 s	2819 s	30 s
(12,3)	529 s	35719 s	36212 s	316 s
(14,4)	No data	No data	No data	12121 s

As shown by the program execution time of the table, the efficiency of PMS algorithms varies greatly with the change of (l,d). By comparison, it shows that Algorithm 1 and PMSP algorithm have similar time and space complexity, however, Algorithm 1 does not have high efficiency in dealing with relatively large values. It should be given enough attention as unlike the other three algorithms that can completely traverse all the motifs, Algorithm 1 may miss out on some of the motifs, which belongs to the local search algorithm. For each input string, a motif instance is added each time, if the total number of motif instances is less than **t**, it is necessary to add **n − l + 1** motif instances to continue depth-first traversal.

By comparing Algorithms 2 and 3, there is no significant difference in execution time, although they differ greatly in the content of the algorithm. They are designed with depth and breadth priority. In the breadth priority since the program needs to store some extra elements in the queue, so inevitably it takes up more memory. Fortunately, memory does not have a significant impact on the execution results of the program, but large memory usage also proves that Algorithm 3 is inefficient.

From the comparison of the above four algorithms, the PMSP algorithm has been proved to have a high efficiency when dealing with the issue of searching motifs, especially when the (l,d) values are relatively large, the other three algorithms have appeared to be unable to obtain operating results. PMSP algorithm does not need to evaluate all l-mers of s_i, but to evaluate the l-mer in the s_i with the Hamming distance of $x \leq 2d$. The candidate motif **y** only needs to be compared with it, resulting in obvious advantages in its execution time.

The PMSP algorithm has a lot of space for optimization. By analyzing the computational characteristics of the PMSP, an algorithm for motif searching based on MapReduce (PMSPMR) is proposed to solve PMS issues. Original problem is evenly distributed to the **P** nodes of the MapReduce distributed system, each node reads all the allocated data and then executes the map function. The scale of the problem to be calculated at each node is **1/P** of original problem, achieving a high degree of parallel computing. After the calculation is completed, corresponding to the results of subproblems are read into a Reduce node, merged the results of the **P** partial questions and added them as solutions to the final output file.

In order to verify the implementation efficiency of PMSPMR algorithm, several representative examples of planted (l,d) motif search issues were selected for experimentation, comparison of algorithm execution time on a single machine and on a 10-node Hadoop cluster (Table 2).

Table 2. Execution time of single machine and 10-node Hadoop cluster algorithms.

(l,d)	Single machine	10-node Hadoop cluster
(10,4)	78 s	33 s
(14,5)	20.6 m	2.6 m
(18,6)	232.2 m	23.9 m

As can be seen from the table that when the single machine runs PMSP algorithm, with the increase of the difficulty of the issue, execution time presents a very obvious increasing trend. For the problem with smaller values of (l,d), the parallel computing performance of the Hadoop cluster is not obvious, the advantages of the Hadoop cluster for the larger (l,d) value are fully reflected and the execution speed is much faster than the standalone execution speed.

5   Conclusions and Future Work

This article mainly introduces and implements related algorithms for planted (l,d) motif search issues. In research process, the method description of the algorithm was first performed and pseudocode was also available. Then in experiment part testing of the algorithms were done, counted and compared the execution time of the programs, explained the reasons for differences and concluded that PMSP algorithm performs best.

Motif searching is a challenging issue that interweaves with many disciplines such as bioinformatics, computational biology and computer science. For the same problem, different motif search algorithms are chosen, which may lead to great difference in program execution efficiency. Researchers should fully consider the time and space complexity of the algorithm when designing the PMS algorithm, this paper combines the theoretical knowledge of MapReduce and uses Hadoop cluster technology to further improve the execution efficiency of the PMSP algorithm. For the planted (l,d) motif search issue, not only PMSP algorithm can achieve distributed parallelization but also other enumeration algorithms, as long as there are good parallel characteristics, can try to use this method to achieve and shorten the problem to solve time. Local search algorithms can also be implemented using this method, which can have a very significant effect on shortening the computation time and improving the accuracy of algorithms.

References

1. Davila, J., Balla, S., Rajasekaran, S.: Space and time efficient algorithms for planted motif search. In: Alexandrov, V.N., van Albada, G.D., Sloot, P.M.A., Dongarra, J. (eds.) ICCS 2006 Part II. LNCS, vol. 3992, pp. 822–829. Springer, Heidelberg (2006). https://doi.org/10.1007/11758525_110
2. Zambelli, F., Pesole, G., Pavesi, G.: Motif discovery and transcription factor binding sites before and after the next-generation sequencing era. Brief Bioinform. 14(2), 225–238 (2013)
3. Mrazek, J.: Finding sequence motifs in prokaryotic Genomes-brief practical guide for a micro-biologist. Brief. Bioinform. 10(5), 525–536 (2009)
4. D'haeseleer, P.: What are DNA sequence motif. Nat. Biotechnol. 24(4), 423–425 (2006)
5. Dean, J., Ghemawat, S.: MapReduce: simplified data processing on large clusters. Commun. ACM 51(1), 107–113 (2008)
6. Hu, J., Li, B., Kihara, D.: Limitations and potentials of current motif discovery algorithms. Nucleic Acids Res. 33(15), 4899–4913 (2005)
7. Rampášek, L., Jimenez, R.M., Lupták, A., et al.: RNA motif search with data-driven element ordering. BMC Bioinform. 17(1), 1–10 (2016)
8. Davila, J., Balla, S., Rajasekaran, S.: Fast and practical algorithms for planted (l,d) motif Search. IEEE/ACM Trans. Comput. Biol. Bioinf. 4(4), 544–552 (2007)
9. Yu, Q., Huo, H., Zhao, R., et al.: RefSelect: a reference sequence selection algorithm for planted (l,d) motif search. BMC Bioinformatics 17(9), 266–281 (2016)
10. He, B., Fang, W., et al.: Mars: a MapReduce framework on graphics processors. In: International Conference on Parallel Architectures and Compilation Techniques, pp. 260–269. IEEE (2017)

11. Hu, J., Li, B., Kihara, D.: Limitations and potentials of current motif discovery algorithms. Nucleic Acids Res. **33**(15), 4899–4913 (2005)
12. Tanaka, S.: Improved Exact Enumerative Algorithms for the Planted(l,d)-Motif Search Problem. IEEE Computer Society Press, Washington, D.C. (2014)
13. Rajasekaran, S., Dinh, H.: A speedup technique for (l,d)-Motif finding algorithms. BMC Res. Notes **4**(1), 1–7 (2011)
14. Peng, X., Pal, S., Rajasekaran, S.: qPMS10: a randomized algorithm for efficiently solving quorum Planted Motif Search problem. In: IEEE International Conference on Bioinformatics and Biomedicine, pp. 670–675. IEEE (2017)
15. Xu, Y., Yang, J., Zhao, Y., et al.: An improved voting algorithm for planted (l,d) motif search. Inf. Sci. **237**, 305–312 (2013)
16. Tong, E., et al.: Bloom filter-based workflow management to enable QoS guarantee in wireless sensor networks. J. Netw. Comput. Appl. **39**, 38–51 (2014)

Knowledge Discovery

Joint Subspace Learning and Sparse Regression for Feature Selection in Kernel Space

Long Chen[1] and Zhi Zhong[2(✉)]

[1] College of Computer and Information Engineering,
Guangxi Teachers Education University, Nanning 530299, China
1063477512@qq.com

[2] College of Continue Education, Guangxi Teachers Education University,
Nanning 530299, China
2823919387@qq.com

Abstract. In this paper, we propose a novel feature selection method to jointly map original data to kernel space and conduct both subspace learning (via locality preserving projection) and feature selection (via a sparsity constraint). The kernel method is used to explore the nonlinear relationship between data and subspace learning is used to maintain the local structure of the data. As a result, we eliminate redundant and irrelevant features and thus make our method select a large amount of informative and distinguishing features. By comparing our proposed method with some state-of-the-art methods, the experimental results showed that the proposed method outperformed the comparisons in terms of clustering task.

Keywords: Feature selection · Kernel method · Subspace learning
Sparse learning · Locality preserving projection

1 Introduction

With the advancement of technology, data collection becomes more and more easy, resulting in an ever-increasing scale and complexity of the database, such as various types of text information, picture information, and biological information, and their features have generally reached hundreds, Thousands or even higher [1]. High-dimensional data not only takes up a lot of storage space and computing time, but also causes a series of problems such as dimension disasters. In addition, not all original features have a positive effect on the results of classification and clustering, and the presence of noise points can affect the accuracy of classification and clustering task. A large number of redundant and irrelevant features will also lead to over-fitting of the learning model, eventually

Supported by organization x.

leading to a weak model generalization ability and loss of utility value. Therefore, the feature reduction of high-dimensional data and mining the important information become a crucial step in machine learning [2].

Feature reduction is to eliminate redundant and irrelevant features through learning model in high-dimensional original data and select a more representative and more accurate new subset. The selected low-dimensional subset is easy to deal with and largely eliminates the interference of noise features, thus making classification and clustering more accurate. Common feature reduction includes two methods: feature selection and subspace learning [3]. Among them, feature selection method selects a subset that best represents the original feature from the original feature space, which can usually be classified into three categories: the filter method [4] first makes feature selection for the dataset and then trains the learner, the feature selection process is independent of the subsequent learner, such as variance selection method; The wrapper method [5] directly takes the performance of the final learner to be used as the evaluation criteria of feature subset, such as recursive feature elimination method; The embedding method [6,7] completes The feature selection and the learner training process in the same optimization process, that is, automatic selection of features during the training of the learner, such as sparse regularization regression. Among them, the effect of embedding method in practice is the most outstanding, mainly due to the local manifold structure is superior to the global structure of data, and embedding method usually try to find the local structure between the data. The subspace learning method is to map high-dimensional space to low-dimensional space by projection [8], so as to ensure the correlation structure between data, such as Locality Preserving Projection (LPP), Principal Component Analysis (PCA), and Linear Discriminant Analysis (LDA), etc.

Although the former feature selection method has good effect in practice, there are at least two shortages as following.

1. The conventional feature selection methods conduct data in the original space, which can only show the shallow hierarchical relationship between data. Data in the real world usually more complex, for example, the features of the main component may be associated with nonlinear input variables, and different categories of data can not be separated by hyperplane, therefore, the conventional method is difficult to excavate the nonlinear relationship between data [9].
2. The form feature reduction method is usually carried out independently by feature selection and subspace learning. Feature selection method can directly select the important features and therefore has a good interpretation. However, subspace learning is more effective than feature selection, but the data after dimension reduction is a linear combination of original features, which have poor explanatory power. Hence, independent implementation of two methods cannot guarantee optimal result when select important features [10].

Research show that the linear inseparable data can be linearly separable if maps it to a kernel space, so using linear feature selection method can achieve the effect of the nonlinear feature selection of original data in kernel space [11]. In order to reduce the influence of the above problems, we propose a novel feature selection method, namely Joint Subspace Learning and Sparse Regression for Feature Selection in Kernel Space (JSLSRK). We highlight the contributions of this paper as follows.

1. In this paper, we map the original data to the kernel space by mapping function, the nonlinear relationship between the data is discovered, so the non-linear correlation problems that are difficult to be solved by previous methods can be solved.
2. We embed the subspace learning method in the sparse constrained feature selection framework, and preserve deeper relationships between data through LPP. Therefore, the proposed algorithm can improve the effect of feature selection and has a good explanatory power.
3. Experiments on the benchmark data set show that the proposed algorithm is more effective than the state-of-the-art feature selection methods.

2 Related Work

In this section, we briefly introduce several popular methods of subspace learning and feature selection, whose models are similar to the algorithms proposed in this paper to some extent. *i.e.* Principle Component Analysis (PCA), Linear Discriminant Analysis (LDA), Laplacian Score for Feature selection (LSFS), Efficient and Robust Feature Selection via Joint $l_{2,1}$-Norm Minimization (RFS), Unsupervised Feature Selection with Structure Graph Optimization (SOGFS), Unsupervised Feature Selection for Multi-Cluster Data (MCFS), and Joint Embedding Learning and Sparse Regression (JELSR). Among them, PCA and LDA are subspace learning methods [12,13], the former is to obtain the eigenvector corresponding to the maximum eigenvalue by calculating the covariance matrix of the data, so as to find the directions with the maximum data variance, and thus achieve the dimensionality reduction effect on the data, and the latter is to project the sample into the new subspace with the largest distance between the class and the smallest within the class distance, in order to obtain the best separability of the sample in the space. PCA is an unsupervised learning method, and LDA is a supervised learning method that needs to use class labels to distinguish sample categories. LSFS, RFS, SOGFS, MCFS, and JELSR are feature selection methods [6,14–17], LSFS method effectively measures the weight of each feature of the sample, and prioritizes those features with less weight. More specifically, by constructing the nearest neighbor graph, the local geometric structure of the data is presented and then the smoothest features on the graph are selected. RFS method constructs a least square regression model, and obtains a weight matrix via employing $l_{2,1}$-norm on both loss function and regularization term. Moreover, the weight of large value corresponds to the important feature, and the weight of small value corresponds to the unimportant feature, so as to achieve the effect of feature selection. SOGFS method gets a global structure through the sample low-dimensional feature space and selects the important features, while uses probability method to construct the similarity matrix between samples and imposes reasonable constraints on it. MCFS method computes the embedding at first and then use regression coefficient to rank each feature, more specifically, first, "unfolding" the data manifold using the manifold learning algorithms, and then put a coefficient constraint on the regression vector, finally, important features would be selected. JELSR method simultaneously takes into account a Laplacian regularizer and the weight matrix to rank the scores of the features.

3   Approach

This paper denote matrices as boldface uppercase letters, vectors as boldface lowercase letters, also denotes the i-th row and j-th column of a matrix $\mathbf{X} = [x_{ij}]$ as xi and x^j, and its Frobenius and l_1-norm as $\|\mathbf{X}\|_F = \sqrt{\sum_i \|x_i\|_2^2}$ and $\|\mathbf{X}\|_1 = \sum_i |x_i|$, and further denotes the transpose, the trace, and the inverse, of a matrix $\mathbf{X}$, as $\mathbf{X}^T$, $tr(\mathbf{X})$ and $\mathbf{X}^{-1}$.

3.1   Sparse Learning

Sparse learning [18] was first applied mainly in the fields of graphics and image visualization [19], because of the strong intrinsic theory and application value, sparse learning obtained fast development, and has been widely used in the field of machine learning and pattern recognition. In the basic theory of sparse learning, first, a sparse assumption is made on the parameter vector $\alpha \in \mathbb{R}^n$ of the model, the training samples are then used to fit the parameters α through the given model. Sparse learning can be expressed as:

$$\min_{\alpha} \ g(\alpha) = f(\alpha) + \lambda\phi(\alpha) \tag{1}$$

Where $f(\alpha)$ is the loss function, $\phi(\alpha)$ is the regularization term, λ is the regularization parameter, is used to adjust the sparsity of α. By applying sparse learning to feature selection algorithm, coefficient weight between sample features can be introduced into the model as important discriminant information. The input data can be expressed sparsely through the sparse constraint, which can eliminate redundant and irrelevant attributes and ensure that important features can be selected. In order to ensure that the proposed algorithm in this paper can obtain the unique global optimal solution, the regularization factor of sparse learning can choose the norm form that can be solved by convex optimization. In sparse learning, l_0-norm is the most effective sparse regular factor, but it is NP hard problem. Moreover, it has been proved in theory that l_1-norm is the optimal convex approximation of l_0-norm, so many literatures use l_1-norm instead of l_0-norm [20].

3.2   Proposed Method

This paper defines the sample matrix $\mathbf{X} \in \mathbb{R}^{n \times d}$, Where the n and d represents the number of samples and the number of features respectively, $\mathbf{Y} \in \mathbb{R}^{n \times m}$

defines class indicator matrix or class label matrix, Where m represents the number of categories. In the case of supervised learning, for any matrix according to the least squares regression model of multitasking learning, combining with the regularization constraint on the reconstruction coefficient, firstly defining the multi-class feature selection problem formula as [21]:

$$\min_{\mathbf{W}} l(\mathbf{Y} - \mathbf{XW}) + \lambda R(\mathbf{W}) \tag{2}$$

Where $\mathbf{W}$ is the feature weight matrix, $l(\mathbf{Y} - \mathbf{XW})$ is the loss function, $R(\mathbf{W})$ is the regular term, λ is the regularization parameter. Although Eq. (2) is a widely used method for feature selection, the non-linear relationship between data cannot be reflected when the original data is input into the loss function. For complex nonlinear correlation data, the model of Eq. (2) is often unable to effectively select important features, so the model with poor generalization ability will be trained. Considering the above effects, a nonlinear attribute selection algorithm is proposed in this paper. First, the original data set is mapped to the kernel space by a linear kernel function, assume that a sample matrix $\mathbf{X} \in \mathbb{R}^{n \times d}$, where $x_i \in \mathbb{R}^{d \times 1} (i = 1, 2, \ldots, n)$, The definition of linear kernel function is $\mathbf{K}(x_i, x_j) = <x_i, x_j>$ [22]. The kernel for each sample is expressed as $\mathbf{K}(x_i, \bullet) = [x_i^T x_1, x_i^T x_2, \ldots, x_i^T x_n]$. So the original sample matrix would have a kernel matrix $\mathbf{K} \in \mathbb{R}^{n \times n}$, where $\mathbf{K}_{ij} = \mathbf{K}(x_i, x_j) = x_i^T x_j$, So the feature selection model in kernel space becomes:

$$\min_{\mathbf{W}} l(\mathbf{Y} - \mathbf{KW}) + \lambda R(\mathbf{W}) \tag{3}$$

Because the sparse of $\mathbf{W}$ cannot eliminate the features in the original data, therefore, the sparse constraint of $\mathbf{W}$ cannot be used to make feature selection of $\mathbf{X}$. This paper use a more ingenious method to implement the feature selection of samples. First, each column in the original data matrix is treated as a new matrix, every new matrix $x^j \in \mathbb{R}^{n \times 1} (j = 1, 2, \ldots, d)$, each new matrix is projected into the kernel space by a linear kernel function. So we can obtain a kernel matrix for each of these columns $\mathbf{K}_j \in \mathbb{R}^{n \times n} (j = 1, 2, \ldots, d)$, there are d kernel matrices. Further, it can be deduced that the feature selection model in the kernel space is:

$$\min_{\mathbf{W}, \alpha} l\left(\mathbf{Y} - \sum_i^d \alpha_i \mathbf{K}^i \mathbf{W}\right) + \lambda_1 R(\mathbf{W}) + \lambda_2 R(\alpha) \tag{4}$$

Where $\alpha \in \mathbb{R}^{d \times 1}$ is the weight matrix of the kernel matrix, which represents the importance of each attribute in the sample, $\mathbf{K}^i$ represents i-th feature's kernel matrix, $\mathbf{W} \in \mathbb{R}^{n \times m}$ is the projection matrix in kernel space. In order to make the loss term in Eq. (4) fit $\mathbf{Y}$ as best as possible, the Frobenius is used to estimate the remainder as $\min_{\mathbf{W}, \alpha} \left\| \mathbf{Y} - \sum_i^d \alpha i \mathbf{K} i \mathbf{W} \right\|_F^2$, meanwhile, a penalty term $R(\mathbf{W})$ is added to prevent model over-fitting, to remove noise and redundant data, continue adding regular term $R(\alpha)$. Because the l_1-norm can lead to sparse structure, some insignificant coefficients in the penalty term will be reduced or directly to 0, which can exclude the unimportant features [23,24], therefore, in this paper, the l_1-norm is used for sparse constraint on α, and the Frobenius is used to take $\mathbf{W}$ as the target term of regularization punishment, *i.e.*

$$\min_{\mathbf{W},\alpha} \frac{1}{2} \left\| \mathbf{Y} - \sum_i^d \alpha_i \mathbf{K}^i \mathbf{W} \right\|_F^2 + \lambda_1 \|\mathbf{W}\|_F^2 + \lambda_2 \|\alpha\|_1 \tag{5}$$

In order to maintain the near-neighbor relationship after spatial projection transformation and further increase the generalization ability of the model, this paper embedded the subspace learning method LPP in the attribute selection model. The main role of LPP algorithm is projecting data to low-dimensional space and preserving neighborhood structure of samples in the space, at the same time, the method of local merging is used to reconstruct the overall internal rules of data. Assuming that we have a data set $\mathbf{X} \in \mathbb{R}^{m \times n}$, where m represents the number of samples and n represents the number of features of samples. If we can find a matrix $\mathbf{A}$ that allows each of samples of matrix $\mathbf{X}$ to be projected into a subspace $\mathbb{R}^d$ $(d \ll n)$, and there is $\mathbf{Y} = \mathbf{A}^T \mathbf{X}$ $(\mathbf{Y} \in \mathbb{R}^{m \times d})$. The concrete steps of the LPP algorithm are as follows.

1. Construct the neighbor graph: each sample $x_i (x_i \in \mathbf{X})$ is treated as a vertex of an undirected graph, and then the k nearest neighbor algorithm $(k \in \mathbf{N})$ is used to determine whether it is a nearest neighbor, if it is a close neighbor, it will be connected by a line, otherwise it will not be connected, and finally forming a close neighbor diagram.
2. Construct the weight matrix: it is defined $\mathbf{S} \in \mathbb{R}^{m \times m}$ as a sparse symmetric matrix, where $\mathbf{S}_{ij}$ represents the weight of edges of vertex x_i and vertex x_j, In the nearest neighbor graph, if the two samples have no edge connection, it indicates that the two samples are independent, then $\mathbf{S}_{ij} = 0$, the two vertices with edge connection are defined by the heat kernel as $\mathbf{S}_{ij} = \exp(-\frac{\|x_i - x_j\|}{t})$, Where t is a constant greater than 0.
3. Feature mapping: the ultimate goal of the algorithm is to find an optimal projection direction, and the criterion for selecting the optimal direction is to minimize the following target function.

$$\frac{1}{2} \sum_{ij} \mathbf{S}_{ij} \|x_i \mathbf{A} - x_j \mathbf{A}\|_2^2 \tag{6}$$

Considering the property of LPP, the following objective functions can be defined in this paper:

$$\frac{1}{2} \sum_i^d \sum_{p,q}^n \mathbf{S}_{pq}^i \|k_p^i \mathbf{W} - k_q^i \mathbf{W}\|_2^2 \tag{7}$$

Where $\mathbf{S} = [\mathbf{S}_{p,q}] \in \mathbb{R}^{n \times n}$ is the similarity matrix, and $H(k_p, k_q) = \exp[-\frac{\|k_p - k_q\|}{t}]$. By simple mathematical transformation, Eq. (7) can be transformed into Eq. (8), *i.e.*

$$\frac{1}{2} \sum_i^d \sum_{p,q}^n \mathbf{S}_{pq}^i \left\| k_p^i \mathbf{W} - k_q^i \mathbf{W} \right\|_2^2$$

$$\Leftrightarrow \sum_i^d [\sum_p^n \mathbf{W} k_p^i \mathbf{D}_{pp}^i (k_p^i)^T \mathbf{W}^T - \sum_{p,q}^n \mathbf{W} k_p^i \mathbf{S}_{pq}^i (k_p^i)^T \mathbf{W}^T]$$

$$\Leftrightarrow \sum_i^d tr[\mathbf{K}^i \mathbf{W} (\mathbf{D}^i - \mathbf{S}^i)(\mathbf{K}^i \mathbf{W})^T]$$

$$\Leftrightarrow \sum_i^d tr[\mathbf{W}^T (\mathbf{K}^i)^T \mathbf{L}^i \mathbf{K}^i \mathbf{W}] \tag{8}$$

Where $\mathbf{D}$ is the diagonal matrix, which its diagonal elements are the sum of each row in the matrix $\mathbf{S}$, $i.e.$ $\mathbf{D}_{pp}^i = \sum_q \mathbf{S}_{pq}^i$. $\mathbf{L}$ is the Laplace matrix and is defined as $\mathbf{L} = \mathbf{D} - \mathbf{S}$. The matrix $\mathbf{D}$ measures the importance of this sample, that is, the bigger the $\mathbf{D}_{pp}$ is, the more important the p-th sample is, and so the y_p after the projection is more important. Finally, Eq. (8) is incorporated into Eq. (5) in this paper, and the objective function of proposed algorithm is obtained. $i.e.$

$$\min_{\mathbf{W},\alpha} \frac{1}{2} \left\| \mathbf{Y} - \sum_i^d \alpha_i \mathbf{K}^i \mathbf{W} \right\|_F^2 + \lambda_1 \|\mathbf{W}\|_F^2 + \lambda_2 \|\alpha\|_1 + \lambda_3 \sum_i^d tr[\mathbf{W}^T (\mathbf{K}^i)^T \mathbf{L}^i \mathbf{K}^i \mathbf{W}]$$
$$\tag{9}$$

Where $\mathbf{Y} \in \mathbb{R}^{n \times m}$ is the label matrix of the sample, $\mathbf{K} \in \mathbb{R}^{n \times n}$ is the kernel matrix for each feature of the training sample, $\mathbf{W} \in \mathbb{R}^{n \times m}$ is the projection matrix in kernel space, $\alpha \in \mathbb{R}^{d \times 1}$ is the weight matrix that represents the importance of features, λ_1, λ_2 and $\lambda 3$ are harmonic parameter. λ_3 is designed to equilibrate the order of magnitude between $\sum_i^d tr[\mathbf{W}^T (\mathbf{K}^i)^T \mathbf{L}^i \mathbf{K}^i \mathbf{W}]$ and $\left\| \mathbf{Y} - \sum_i^d \alpha_i \mathbf{K}^i \mathbf{W} \right\|_F^2$, which the bigger the value of λ_3, the greater the contribution of LPP in Eq. (9) and vice versa [25, 26].

4 Experiment Analysis

In this section, we evaluate our proposed JSLSRK with the comparison methods in terms of the clustering accuracy of the clustering tasks, on six public data sets, whose detail is listed in Table 1. The comparison methods include Laplacian Score for Feature Selection (LSFS), efficient and Robust Feature Selection via Joint $l_{2,1}$-Norms Minimization (RFS), Structured Optimal Graph Feature Selection (SOGFS). Unsupervised Feature Selection for Multi-Cluster Data (MCFS), Joint Embedding Learning and Sparse Regression (JELSR), and baseline which uses all features to conduct k means clustering. In our experiments, we set the

Table 1. Data set description

Data sets	Samples	Features	Class
Corel	1000	423	10
Yale	165	1024	15
Uspst	2007	256	10
Ecoli	336	343	8
Colon	62	2000	2
WARP	130	2400	10

parameters' range as $\{10^{-3}, 10^{-2}, \ldots, 10^3\}$, where all the methods can achieve their best results. We first use all the feature selection methods to select features (*i.e.* $\{10\%, 20\%, \ldots, 90\%\}$ of all the features) and then conduct k means clustering on the selected features. We repeat k means clustering 10 times to report their average results. Finally, we employ the clustering accuracy to evaluate the clustering performance of all the methods.

4.1   Cluster Accuracy

We list the clustering accuracy of all the methods with different numbers of selected features in Fig. 1. Our proposed JSLSRK achieves the best clustering performance, followed by LSFS, RFS, SOGFS, MCFS, JELSR, and baseline. For example, our method on average improves by 8.27% and 2.97%, compared to Baseline (the worst comparison method) and MCFS (the best comparison method). The reason may be that our method (1) effectively excavates the non-linear relationship between the data and effectively selects the features with large amount of information. (2) utilizes locality preversing projection method to maintain the data local manifold structure. We can see from Fig. 2 that the clustering accuracy of all feature selection methods has been improved when selecting useful features, more specifically, the LSFS method (the worst feature selection method) improved the clustering accuracy of 3.92% compared to the baseline. Therefore, it is shown that the removal of redundant and irrelevant features and the selection of important features have a positive effect on clustering tasks. Moreover, in Fig. 2, some clustering accuracy will be improved with the increase of selection features (such as Yale dataset), that means that most of the features of the data set are important and good for clustering tasks. And the accuracy of other clustering decreases with more features selected (such as WARP and Colon datasets), reveals that most of the features are invalid, and only a small number of features determines the label of samples. In addition to above situation, some clustering accuracy initially increases with the increase of the selected features, but gradually decreases or remains unchanged after reaching a maximum (such as Ecoli and Uspst datasets). This trend indicates that a small number of important features and an excessive number of redundant features could output bad clustering performance.

4.2   Parameters Sensitivity and Convergence

Our objective function has three tuning parameters, *i.e.*, λ_1, λ_2 and λ_3, We fix the value of λ_1 and λ_3. In Eq.(9), λ_2 is designed to adjust the sparsity of weight matrix α, the larger the value of λ_2, the more the sparsity of α (*i.e.*, the less features are selected to conduct the clustering tasks). Figure 3 shows the highest accuracy of clustering tasks on each data set when λ_2 takes different values, for example, we can find that our method achieves the best performance on the data sets Uspst and Yale while setting $\lambda_2 = 0.1$ and $\lambda_2 = 1$ respectively [21]. Figure 3 shows the objective function value of each iteration in algorithm 2. In the experiment in this paper, the iteration stopping standard of algorithm 1 and algorithm 2 is set as $(Y_{t+1} - Y_t)/Y_t \leq 10^{-3}$ [27], where Y_t represents the t-th iteration objective function value of Eq. (9). As can be seen from Fig. 3: (1) for the optimization of the proposed objective function Eq. (9), algorithm 2 in this paper will reduce the value of the objective function monotonously until the function converges. (2) the value of the objective function will converge within 20 times in algorithm 2, which is very efficient. In addition, the algorithm 1 in this paper generally converges within 40 iterations. Due to spatial reasons, it will not be listed one by one.

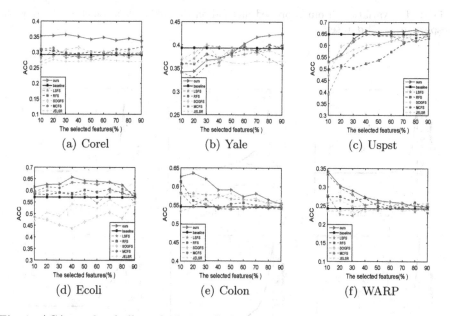

Fig. 1. ACA result of all methods on all data sets at different number of selected features

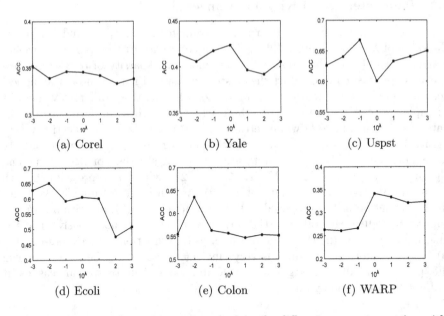

Fig. 2. The variation of our proposed method on the different parameter setting with respect to on all datasets

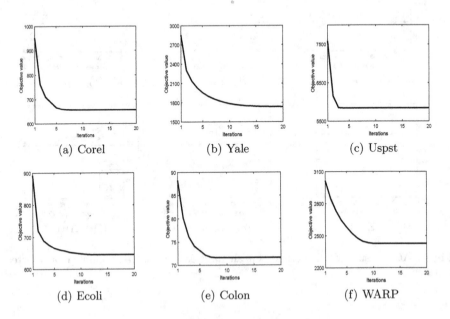

Fig. 3. The convergence of the objective function in Eq. (9) on all datasets.

5   Conclusion

This paper has proposed a novel feature selection method by mapping the data from the original space to the kernel space to explore the nonlinear relationship between the data. Moreover, the l_1-norm regular term is used for sparse constraint to ensure that important attributes can be selected, Finally, the local reservation projection method is embedded to maintain the local structure, which strengthen the ability of attribute selection, and eliminate redundant and irrelevant features. The algorithm effectively integrates attribute selection method and subspace learning method in kernel space, which not only expands the theoretical application of subspace learning, but also makes up for the deficiency of sparse learning in local information processing. The experimental results show that the algorithm can improve the clustering accuracy significantly and converge to the global optimal solution quickly.

References

1. Zheng, W., Zhu, X., Wen, G., Zhu, Y., Yu, H., Gan, J.: Unsupervised feature selection by self-paced learning regularization. Pattern Recogn. Lett. (2018). https://doi.org/10.1016/j.patrec.2018.06.029
2. Zhu, X., Zhang, S., Hu, R., Zhu, Y., et al.: Local and global structure preservation for robust unsupervised spectral feature selection. IEEE Trans. Knowl. Data Eng. **30**(3) pp. 517–529
3. Li, Y., Zhang, J., Yang, L., Zhu, X., Zhang, S., Fang, Y.: Low-rank sparse subspace for spectral clustering. IEEE Trans. Knowl. Data Eng. https://doi.org/10.1109/TKDE.2018.2858782
4. He, X., Cai, D., Niyogi, P.: Laplacian score for feature selection. In: International Conference on Neural Information Processing Systems, pp. 507–514 (2005)
5. Tabakhi, S., Moradi, P., Akhlaghian, F.: An unsupervised feature selection algorithm based on ant colony optimization. Eng. Appl. Artif. Intell. **32**(6), 112–123 (2014)
6. Cai, D., Zhang, C., He, X.: Unsupervised feature selection for multi-cluster data. In: ACM SIGKDD International Conference on Knowledge Discovery and Data Mining, pp. 333–342 (2010)
7. Zhu, X., Li, X., Zhang, S., Xu, Z., Yu, L., Wang, C.: Graph PCA hashing for similarity search. IEEE Trans. Multimed. **19**(9), 2033–2044 (2017)
8. Zhang, S., Li, X., Zong, M., Zhu, X., Wang, R.: Efficient kNN classification with different numbers of nearest neighbors. IEEE Trans. Neural Netw. Learn. Syst. **29**(5), 1774–1785 (2018)
9. Cao, B., Shen, D., Sun, J.T., Yang, Q., Chen, Z.: Feature selection in a kernel space. In: Proceedings of the Twenty-Fourth International Conference on Machine Learning, pp. 121–128 (2007)
10. Hu, R., et al.: Graph self-representation method for unsupervised feature selection. Neurocomputing **220**, 130–137 (2017)
11. Baudat, G., Anouar, F.: Generalized discriminant analysis using a kernel approach. Neural Comput. **12**(10), 2385–2404 (2000)
12. Zhi, X., Yan, H., Fan, J., Zheng, S.: Efficient discriminative clustering via QR decomposition-based linear discriminant analysis. Knowl.-Based Syst. **153**, 117–132 (2018)

13. Rahmani, M., Atia, G.K.: Coherence pursuit: fast, simple, and robust principal component analysis. IEEE Trans. Sig. Process. **65**(23), 6260–6275 (2017)

14. Zhu, L., Miao, L., Zhang, D.: Iterative Laplacian score for feature selection. In: Liu, C.-L., Zhang, C., Wang, L. (eds.) CCPR 2012. CCIS, vol. 321, pp. 80–87. Springer, Heidelberg (2012). https://doi.org/10.1007/978-3-642-33506-8_11

15. Nie, F., Huang, H., Cai, X., Ding, C.: Efficient and robust feature selection via joint $\ell_{2,1}$-norms minimization. In: International Conference on Neural Information Processing Systems, pp. 1813–1821 (2010)

16. Nie, F., Zhu, W., Li, X.: Unsupervised feature selection with structured graph optimization. In: Thirtieth AAAI Conference on Artificial Intelligence, pp. 1302–1308 (2016)

17. Hou, C., Nie, F., Li, X., Yi, D., Wu, Y.: Joint embedding learning and sparse regression: a framework for unsupervised feature selection. IEEE Trans. Cybern. **44**(6), 793–804 (2014)

18. Lai, H., Pan, Y., Liu, C., Lin, L., Wu, J.: Sparse learning-to-rank via an efficient primal-dual algorithm. IEEE Trans. Comput. **62**(6), 1221–1233 (2013)

19. Zhu, X., Li, X., Zhang, S., Ju, C., Wu, X.: Robust joint graph sparse coding for unsupervised spectral feature selection. IEEE Trans. Neural Netw. Learn. Syst. **28**(6), 1263–1275 (2017)

20. Alamri, A.A.: Theory and methodology on the global optimal solution to a General Reverse Logistics Inventory Model for deteriorating items and time-varying rates. Comput. Ind. Eng. **60**(2), 236–247 (2011)

21. Zheng, W., Zhu, X., Zhu, Y., Hu, R., Lei, C.: Dynamic graph learning for spectral feature selection. Multimed. Tools Appl. (2017). https://doi.org/10.1007/s11042-017-5272-y

22. Baudat, G., Anouar, F.: Feature vector selection and projection using kernels. Neurocomputing **55**(1), 21–38 (2003)

23. Zhu, X., Zhu, Y., Zhang, S., Hu, R., He, W.: Adaptive hypergraph learning for unsupervised feature selection. In: Twenty-Sixth International Joint Conference on Artificial Intelligence, pp. 3581–3587 (2017)

24. Zhu, X., Zhang, S., Hu, R., Zhu, Y., Song, J.: Local and global structure preservation for robust unsupervised spectral feature selection. IEEE Trans. Knowl. Data Eng. **30**(3), 517–529 (2018)

25. Gu, Q., Li, Z., Han, J.: Linear discriminant dimensionality reduction. In: Gunopulos, D., Hofmann, T., Malerba, D., Vazirgiannis, M. (eds.) ECML PKDD 2011. LNCS (LNAI), vol. 6911, pp. 549–564. Springer, Heidelberg (2011). https://doi.org/10.1007/978-3-642-23780-5_45

26. Zhu, X., Zhang, L., Huang, Z.: A sparse embedding and least variance encoding approach to hashing. IEEE Trans. Image Process. **23**(9), 3737–3750 (2014)

27. Zhu, X., Suk, H.-I., Huang, H., Shen, D.: Low-rank graph-regularized structured sparse regression for identifying genetic biomarkers. IEEE Trans. Big Data **3**(4), 405–414 (2017)

Adaptive Graph Learning for Supervised Low-Rank Spectral Feature Selection

Zhi Zhong[✉]

College of Continuing Education, Guangxi Teachers Education University,
Nanning 530001, Guangxi, People's Republic of China
2823919387@qq.com

Abstract. Spectral feature selection (SFS) is getting more and more attention in recent years. However, conventional SFS has some weaknesses that may corrupt the performance of feature selection, since (1) SFS generally preserves the either global structure or local structure, which can't provide comprehensive information for the model; (2) graph learning and feature selection of SFS is two individual processes, which is hard to achieve the global optimization. Thus, a novel SFS is proposed via introducing a low-rank constraint for capturing inherent structure of data, and utilizing an adaptive graph learning to couple the graph learning and feature data learning in an iterative framework to output a robust and accurate learning model. A optimization algorithm is proposed to solve the proposed problem with a fast convergence. By comparing to some classical and first-class feature selection methods, our method has exhibited a competitive performance.

Keywords: Low-rank constraint · Spectral feature selection
Adaptive graph learning

1 Introduction

Data analysis urgently needs improvement of precision in the learning model, which lead to the high-dimensional data is becoming a necessary precondition, since it can provide more useful information for the learning model, so that model can yield a more accurate result. However, for a specific tasks, the important information is usually preserved in part features rather than all features [13]. Thus, high-dimensional data generally contains a number of redundant features. For the redundancy, it not only increase the storage space, speed up the execution time, but also corrupt the performance of learning model. More important, the over high dimension may lead to the failure of model construction because of the curse of dimensionality. Thus, it motivates us to decrease the redundancy of data as well as preserve useful information.

Dimensionality reduction method is used to decrease the dimensionality of data, which has been developed into an independent and important subject in the

© Springer Nature Singapore Pte Ltd. 2018
Q. Chen et al. (Eds.): ATIS 2018, CCIS 950, pp. 159–171, 2018.
https://doi.org/10.1007/978-981-13-2907-4_14

machine learning, patter recognition, and so on [14]. The method of dimensionality reduction has different categories in different perspectives. Firstly, It can be partitioned into the unsupervised dimensionality reduction learning and supervised dimensionality reduction learning based on the usage of the label information [11]. Specifically, unsupervised dimensionality reduction is designed to preserve the global/local correlation of data, or keep the distribution of data won't be taken changed. By contrast, supervised dimensionality reduction is designed to take advantage of the label information to drive the process of dimensionality reduction. Besides, the dimensionality reduction method also can be partitioned into the feature selection and subspace learning depending on property of feature in the dimensionality-reduced data. More specifically, subspace learning is to find a reasonable projection model to present the correlation or distribution (such as the local or local structure of data) of original high-dimensional data by a new subspace data. The classical subspace learning methods includes the Principal Component Analysis (PCA), Locality Preserving Projections (LPP), Canonical Correlation Analysis (CCA), and so on. There also has the corresponding kernel form to process the nonlinear data set [16]. Feature selection is a interpretable method to directly delete the unimportant and irrelative features to preserve the informative features in the subset. In summary, subspace learning is a robust model that easily output a dimensionality-reduced subset. However, it has changed the original features and thus miss the interpretability. By contrast, feature selection is an interpretable selection model with limited robustness. In real application, interpretable data (*i.e.,* consist of the original features) is convenient to the research task, which makes feature selection task is more practice by comparing to subspace learning methods.

In order to design a dimensionality reduction learning model with both robustness and interpretability, spectral feature is thus designed by embedding the subspace leaning into the feature selection framework, via taking use of the graph theory to store the data's correlation [15,23]. By comparing to the conventional feature selection learning model, the SFS generally can yield a better performance since it preserves the local or global information of data. Although spectral method has an obvious improvement, it still has some weaknesses that may corrupts the performance of feature selection. Firstly, conventional SFS usually take the either local structure or global structure preservation into account. However, this method can't provide comprehensive information for the model training [5,19]. Secondly, conventional SFS generally firstly construct a fixed graph matrix to preserve the correlation of data, and then take use of this graph matrix to conduct the feature selection [15]. As we all know, original feature data usually has redundancy and noise, which obviously corrupt the quality of the graph matrix, and the process of SFS tightly relates with the graph matrix, *i.e.,* a corrupted graph matrix easily result in a bad results. Moreover, conventional SFS usually separates these two steps (*i.e.,* graph matrix construction and feature data learning) into two individual processes, which although can lead to the individual optimization of two steps, it is hard to achieve the final global optimization of feature selection.

As to above three problems, we in this paper have proposed a novel supervised feature selection algorithm by referring to a low-rand constraint and an adaptive graph learning. For our proposed method, we push an effective contrast on the weight matrix to limit its rank, and further embeds a graph learning into the learning framework to simultaneously consider about the local structure preservation and global structure preservation in our proposed model. Hence, we can provide comprehensive information for the learning model to output an informative subset. Besides, we further take advantage of the adaptive graph learning to integrate the graph matrix construction and feature data learning to make the graph matrix constructed from a pure subset, and feature selection can rely on a high-quality graph matrix. By iterations, our proposed feature selection algorithm can achieve its global optimization.

Three main contributions of our paper are summarised in follow:

- Firstly, the proposed method simultaneously consider about the local structure preservation and global structure preservation into a same learning model, which can take full use of the comprehensive information to enhance the learning ability of model to output an informative and representative subset. By comparing to the conventional methods, such as CSFS [2] that only consider about the global structure preservation, and the LPP that separately considers about the local structure preservation, our proposed method can easily output a more robust performance.
- Secondly, our algorithm has employed an adaptive graph matrix rather than a fixed graph matrix to serve for the process of feature selection, so that the model can learn a graph matrix from the pure subset rather than the original feature data. By comparing to the methods such as in [24] that constructs a fixed graph matrix from the data with redundancy, our method iteratively optimizes the graph matrix construction and feature data learning when the results of both steps keep stable. By this way, our method can construct a more accurate graph to reflect the inherent correlation of data, and eventually yield a better performance.
- Thirdly, we also proposed a novel optimization algorithm to speed up the convergence of the objective problem. Besides, extensive experiments on real data sets have exhibited our method outperforms the comparisons.

2 Related Work

2.1 Notations

Through all the paper, we denote matrices, vectors, scale respectively as boldface uppercase letters, boldface lowercase letters, and normal italic letters. Given a matrix $\mathbf{M}$, we use the $\mathbf{m}_{i,\cdot}$, $\mathbf{m}_{\cdot,j}$ and $m_{i,j}$ to respectively represent the i-th row, j-th column, and the element in the i-th row and j-th column. Besides, we denote the Frobenius norm and $\ell_{2,1}$-norm of a matrix $\mathbf{M}$ as $||\mathbf{M}||_F = \sqrt{\sum_{i,j} \mathbf{m}_{i,j}^2}$, and $||\mathbf{M}||_{2,1} = \sum_i \sqrt{\sum_j m_{ij}^2}$, respectively. We further denote the rank operator, the transpose operator, the trace operator, and the inverse of a matrix $\mathbf{M}$ as $rank(\mathbf{M})$, $\mathbf{M}^T$, $tr(\mathbf{M})$, and $\mathbf{M}^{-1}$, respectively [23].

2.2 Feature Selection Based on Sparsity Learning

Sparsity learning model is a kind of popular feature selection model, via introducing the sparsity regularization to penalize the weight matrix to be sparse, so that it can recognize important features [18, 20]. The sparsity learning based feature selection framework can be represented by formula as:

$$\min_{\mathbf{W}} \eta(\mathbf{W}) + \rho(\mathbf{W}), \tag{1}$$

where $\eta(\mathbf{W})$ is a regression function and $\rho(\mathbf{W})$ is a sparse regularization on weight matrix. In this paper, we have employed the $\ell_{2,1}$-norm to conduct sparse result, as it can result in a result with row-sparsity that pushing the row of weight matrix corresponding to the unimportant features to be zero [21]. In order to take advantage of the label information of data, we have employed the least square loss function as regression function in our model. Thus, our basic feature selection framework based on the sparsity learning can be changed into:

$$\min_{\mathbf{W}} \|\mathbf{Y} - \mathbf{XW}\|_F^2 + \alpha\|\mathbf{W}\|_{2,1}, \tag{2}$$

where $\mathbf{Y} \in \mathbb{R}^{n \times c}$ is the label information of data, $\mathbf{X} \in \mathbb{R}^{n \times d}$ is the feature data, moreover, n, c, d respectively donates the number of sample, class, and feature. Besides, the tuning parameter α is used for balancing the magnitude of two terms of Eq. 2, *i.e.*, the regression term and sparsity term.

2.3 Global Structure Preservation Based on Low-Rank Constraint

Equation (2) learns a sparse representation to conduct the feature selection task. However, it does not take any correlation of data into account, which obviously depress the performance of learning model. For our method, we suppose to keep the global information of original data in the selected subset. As we know, real data usually consists of more or less redundant information and noise. They absolutely increase the rank of the matrix of data. Motivated by the work [9, 17], we know that the low-rank constraint can effectively relieve the affection of noise to capture the inherent structure of data, so that eventually output a robust feature selection model. For the Eq. (2), assume that there has a low-rank constraint on the weight matrix $\mathbf{W}$, that is, $\mathbf{W} = \mathbf{PR}$, where $\mathbf{P} \in \mathbb{R}^{d \times r}$, $\mathbf{R} \in \mathbb{R}^{r \times c}$ and $r \leq min(d, c)$. Equation (2) can be rewritten as:

$$\min_{\mathbf{PR}} \|\mathbf{Y} - \mathbf{XPR}\|_F^2 + \alpha\|\mathbf{PR}\|_{2,1}. \tag{3}$$

For Eq. (3), $\mathbf{P}$ is regarded as an projection matrix that projects the original data into a new feature subspace spanned by $\mathbf{XP}$. Actually, it is exactly the optimal subspace projected by the LDA that considers about the global information, which has been proved by theoretical analysis and experiments in [1]. And then, another projection matrix $\mathbf{R}$ is employed to minimize regression error between the new projected feature space and the label information. By this way, we can guarantee that the projected feature data has obvious global structure as same as the structure of original data.

2.4 Adaptive Graph Learning for Local Structure Preservation

Spectral feature selection is one of important branches of feature selection method. It usually embeds the subspace learning in feature selection framework to maintain the local or global manifold structure in the feature-selected subset. Specifically, the LPP is widely used to make the locality preserving projections can be performed in the feature selection method [12,22]. Assuming that there has two similar data points $\mathbf{x}_{i,\cdot}$ and $\mathbf{x}_{j,\cdot}$ in original feature space, they will keep the similar relationship in the projected feature subspace, which can be formulated as:

$$\min \mathbf{W} \sum_{i,j}^{n} \|\mathbf{x}_{i,\cdot} \mathbf{W} - \mathbf{x}_{j,\cdot} \mathbf{W}\|_2^2 s_{i,j}, \tag{4}$$

where $\mathbf{W}$ is the projection matrix (*i.e.*, the weight matrix) that projects original data (*i.e.*, $\mathbf{X}$) into a new feature space (*i.e.*, $\mathbf{XW}$), and $\mathbf{S}$ is a similarity matrix preserving the correlation of original data [10]. And this similarity matrix is usually constructed by k-nearest neighbors strategy by:

$$s_{i,j} = f(\mathbf{x}_{i,\cdot}, \mathbf{x}_{j,\cdot}) = exp(-\frac{\|\mathbf{x}_{i,\cdot} - \mathbf{x}_{j,\cdot}\|_2^2}{2\sigma^2}). \tag{5}$$

Although the LPP constraint (*i.e.*, Eq. (4)) is widely embedded into the spectral feature selection model and has achieved obvious improvement, it exists a obvious weaknesses. As we know, we generally use the original data to construct this similarity matrix. However the original data usually contains redundancy and noise. Thus, the constructed similarity matrix can't effectively preserve the local correlation of data, which directly depress the performance of feature selection. To resolve this problem, it demand us to perform feature selection and graph matrix construction simultaneously. By referring to wok [6], we can have:

$$\min_{\mathbf{S},\mathbf{W}} \sum_{i,j}^{n} (\|\mathbf{x}_{i,\cdot} \mathbf{W} - \mathbf{x}_{j,\cdot} \mathbf{W}\|_2^2 s_{i,j} + \beta \|\mathbf{s}_{i,j}\|_2^2), \\ s.t., \forall i, s_i^T \mathbf{1} = 1, \mathbf{s}_{i,i} = 0 \tag{6}$$

where there is a neighbour of i-th samples, there is a nonnegative $s_{i,j}$, otherwise, the $s_{i,j}$ is set as zero, that is: $s_{i,j} \geq 0$ *if* $i \in N(j)$ ($N(j)$ stands for all nearest neighbours of j-th sample), *otherwise* 0. We have used a parameter β to balance two terms of Eq. (6). For the formula, feature selection and graph matrix learning (*i.e.*, $\mathbf{W}$ and $\mathbf{S}$) can be adaptively updated until achieving the ultimate global optimization. As a result, the optimal graph matrix is constructed by the feature-selected subset, which then is used to conduct the final result of feature selection.

3   Proposed Method

In the section, we intend to design a novel feature selection framework by simultaneously considering about the local structure preservation and global structure preservation in a unified framework, respectively though a low-rank constraint and an adaptive graph matrix learning. Thus, we suppose to embed the low-rank constraint (*i.e.*, Sect. 2.3) and adaptive graph matrix learning (*i.e.*, Sect. 2.4) into the basic feature selection framework based on the sparsity learning (*i.e.*, Sect. 2.2), to output our final objective equation, that is:

$$\min_{S,P,R} ||Y - XPR||_F^2 + \alpha \sum_{i,j}^n (||x_{i,\cdot}PR - x_{j,\cdot}PR||_2^2 s_{i,j} + \beta s_{i,j}^2) + \gamma ||PR||_{2,1}$$
$$s.t., \forall i, s_i^T 1 = 1, s_{i,i} = 0, s_{i,j} \geq 0 \ if \ i \in N(j) \quad otherwise \ 0, \qquad (7)$$
$$rank(PR) = r, where \ r \leq min(d,c)$$

For the Eq. (7), we first use the least square lost function to take use the label information to assist the guidance of the feature selection, and then embed the $\ell_{2,1}$-norm regularizer to conduct the sparsity learning. It can eventually perform the feature selection. Besides, with the purpose improving the performance of learning model to accurately select informative features, we further introduce two subspace learning methods into the feature selection model to output a novel dimensionality reduction model with both robustness and interpretability. Meanwhile, one is the LDA via low-rank constraint, and another method is LPP via graph matrix learning. Specifically, low-rank constraint can effectively find the adaptive subspace structure from the original feature data with redundancy and perform the global structure preservation, which actually has been proved to perform the approximate LDA subspace projection. We will prove it in the following section. Different from classical LPP embedded in conventional spectral feature selection to maintain the local information, we have modified it into an adaptive form. That is, our method can adaptively learn an accurate graph matrix from a dimensionality-reduced data, so that our method can learn an accurate graph matrix from original feature data with redundant features. It means that our method can iteratively update the graph matrix construction and feature selection process, so that model can effectively resolve the problem that redundancy corrupts the quality of constructed graph matrix. By this way, it can finally output a more robust performance of feature selection.

4   Optimization

For the objective function (*i.e.*, Eq. (7)), there has total three variables (*i.e.*, **S**, **P**, and **R**) need to optimize. Although it not joint convex to all variables, it is convex to individual variable when other variables is fixed. Thus, it supposes to use the alternative strategy to optimize the objective function, *i.e.*, (1) update **P** and **R** by fixing **S**; (2) update **S** by fixing **P** and **R**. The optimization process is simplified in 1.

Algorithm 1. The process of optimization algorithm to solve Eq. (7).

Input: $\mathbf{X} \in \mathbb{R}^{n \times d}$, $\mathbf{Y} \in \mathbb{R}^{n \times c}$, α, γ, k, and r;
Output: $\mathbf{P} \in \mathbb{R}^{n \times r}$, $\mathbf{R} \in \mathbb{R}^{r \times c}$, $\mathbf{S} \in \mathbb{R}^{n \times n}$;
1. initialize $\mathbf{S}$ by Eq. (5);
2. initialize β by Eq.(16);
3. **repeat:**
  3.1 **repeat:**
    3.1.1 Update $\mathbf{R}$ by Eq. (11);
    3.1.2 Update $\mathbf{P}$ by optimizing the Eq.(12);
    3.1.2 Update $\mathbf{H}$ by Eq.(10);
    until Eq. (8) converges
  3.2 Update $\mathbf{S}$ by Eq.(15);
until Eq. (7) converges

4.1 Update P and R by Fixing S

By fixing the $\mathbf{S}$, the objective equation (*i.e.*, Eq. (7)) can be rewritten as:

$$\min_{\mathbf{P},\mathbf{R}} \|\mathbf{Y} - \mathbf{XPR}\|_F^2 + \alpha \sum_{i,j}^{n} \|x_{i,.}\mathbf{PR} - x_{j,.}\mathbf{PR}\|_2^2 s_{i,j} + \gamma \|\mathbf{PR}\|_{2,1}. \tag{8}$$

For this formulation, the $\ell_{2,1}$-norm regularizer make the function non-smooth. Thus, the framework of Iteratively Reweighted Least Square (IRLS) can be used to optimize this problem [4]. That is:

$$\min_{\mathbf{P},\mathbf{R}} \|\mathbf{Y} - \mathbf{XPR}\|_F^2 + \alpha tr(\mathbf{R}^T\mathbf{P}^T\mathbf{X}^T\mathbf{LXPR}) + \gamma tr(\mathbf{R}^T\mathbf{P}^T\mathbf{HPR}), \tag{9}$$

where $\mathbf{L}$ is defined as a Laplacian matrix represented by $\mathbf{L} = \mathbf{Q} - \mathbf{S}$, and $\mathbf{Q}$ is a diagonal matrix. The element i-th diagonal element of matrix $\mathbf{Q}$ can be defined as $q_{i,i} = \sum_{j=1}^{n} s_{i,j}$. For the formula, the matrix $\mathbf{H}$ is also a diagonal matrix which i-th element can be denoted as:

$$h_{ii} = \frac{1}{2\|(\mathbf{PR})_{i,.}\|_2^2}, i = 1, \ldots, d, \tag{10}$$

By fixing the $\mathbf{R}$ and making the derivative of Eq. (9) to be zero by respecting to $\mathbf{R}$, it is easy to work out the $\mathbf{P}$, that is:

$$\mathbf{R} = (\mathbf{P}^T\mathbf{S}_t\mathbf{P})^{-1}\mathbf{Y}^T\mathbf{XP}, \tag{11}$$

where $\mathbf{S}_t = \mathbf{XX}^T + \alpha\mathbf{X}^T\mathbf{LX} + \gamma\mathbf{H}$. After working out the $\mathbf{R}$ and making the derivative of Eq. (9) to be zero by respecting to $\mathbf{P}$, we thus can work out the $\mathbf{R}$, under the precondition that substitutes the Eq. (11) into Eq. (9). We can have following expression:

$$\max_{\mathbf{P}} tr(\mathbf{P}^T\mathbf{S}_t\mathbf{P})^{-1}\mathbf{P}^T\mathbf{S}_b\mathbf{P}, \tag{12}$$

where $\mathbf{S}_b = \mathbf{X}^T \mathbf{Y} \mathbf{Y}^T \mathbf{X}$. By comparing to the LDA, $\mathbf{S}_t$ and $\mathbf{S}_b$ can be respectively similarly regarded as within-class scatter matrix and between-class scatter matrix. Meanwhile, the matrix $\mathbf{P}$ is the projection matrix performing the similar projection function in LDA method. Thus, low-rank constraint actually perform the subspace projection in the LDA space. Thus, it is easy to work out the Eq. (12), *i.e.*, the top r eigenvectors of the $\mathbf{S}_t^{-1} \mathbf{S}_b$. The low-rank constraint projects all subjects into a subspace that can distinguish the class of data, which obviously preserves the global structure.

4.2   Update S by Fixing P and R

After fixing the $\mathbf{P}$ and $\mathbf{R}$, which thus can be regard as a global weight matrix (*i.e.*, $\hat{\mathbf{W}} = \mathbf{PR}$). So, the objective equation is equal to solving:

$$\min_{\mathbf{S}} \alpha \sum_{i,j}^{n} (||\mathbf{x}_{i,.}\hat{\mathbf{W}} - \mathbf{x}_{j,.}\hat{\mathbf{W}}||_2^2 s_{i,j} + \beta s_{i,j}^2) \tag{13}$$
$$s.t., \forall i, \mathbf{s}_i^T \mathbf{1} = 1, s_{i,i} = 0, s_{i,j} \geq 0 \ \textit{if } i \in \mathbf{N}(j) \quad \textit{otherwise } 0$$

As to this formulation, we initialize the $\mathbf{S}$ by the conventional k-nearest neighbour strategy and then gradually update this variable until it achieves its optimization. By denoting $f_{i,j} = \alpha ||\mathbf{x}_{i,.}\hat{\mathbf{W}} - \mathbf{x}_{j,.}\hat{\mathbf{W}}||_2^2$, we can have following expression:

$$\min_{\mathbf{S}} ||\mathbf{s}_i + \frac{1}{2\beta}\mathbf{f}_i||_2^2 \quad s.t., \forall i, \mathbf{s}_i^T \mathbf{1} = 1, s_{i,i} = 0, s_{i,j} \geq 0 \ \textit{if } i \in \mathbf{N}(j) \quad \textit{otherwise } 0, \tag{14}$$

By referring to the Karush-Kuhn-Tucker conditions, we can easily obtain the solution of Eq. (14), that is:

$$s_{i,j} = (-\frac{1}{2\beta}\mathbf{f}_i + \tau)_+, \tag{15}$$

where τ is a Lagrangian multiplier. By this way, the optimization of the objective equation can be worked out.

There has total three tuning parameters in our learning model, which is time consuming to adjust a suitable parameter combination by heuristic strategy to output a robust model. By referring to the work [6], we can know the parameter β in adaptive graph learning can be predefined by following equation:

$$\beta = \frac{k}{2}\hat{f}_{i,k+1} - \frac{1}{2}\sum_{v=1}^{k}\hat{f}_{i,v}, \tag{16}$$

where $\hat{f}$ is the ranked $\mathbf{f}$ by descend order. By this formula, we can reduce one tuning parameter, which effectively reduce the complexity of the proposed learning model.

5   Experiments

In our paper, we perform experiment by four comparison methods and our proposed method on four real data sets and further use classification accuracy to evaluate the performance of feature selection.

5.1 Data Sets and Comparison Methods

We use total four data sets in our experiment. These data sets are downloaded from the UCI Machine Learning Repository[1], which detail situation is showed in the Table 1.

Table 1. The situation of data set.

Data sets	Sample number	Feature number	Class number
Ecoli	330	343	8
Pcmac	1743	3289	2
Uspst	2007	256	10
Yeast	1484	1470	10

Four classical and latest feature selection methods are chosen as comparison methods, which contains a Baseline that directly perform the support vector machine (SVM) task on the original data (*i.e.,* all features are preserved) to evaluate the performance of all feature selection method, a method is designed to preserve the global correlation of data (*i.e.,* CSFS [3]), a method that considers about the local and global correlation (RUFS [8]), and a method that utilizes the adaptive graph learning (*i.e.,* SOGFS [7]).

5.2 Experimental Setting

10-fold cross-validation strategy is used in our paper to conduct all experiments, in order to take full use of all samples. Specifically, the used data set is randomly divided into ten parts on samples. Meanwhile, one in ten samples are used to test the performance of the trained model, and the nine in ten samples are used to train the proposed model. Furthermore, 5-fold cross-validation is further employed to select the best parameters' combination for model construction. For our model, we set parameter k as 10, set the range of remainder both parameters of α and γ as $\{10^{-3}, \ldots, 10^3\}$. We feed the selected subset into SVM to perform the classification task, which ACCuracy (**ACC**) is used to evaluate the performance of feature selection methods. Besides, we also investigate parameters' sensitivity and the convergence of Algorithm 1 to test the robustness and efficiency of our method and optimization algorithm.

5.3 Experimental Result

We list the results of feature selection with different number of preserved feature in the Fig. 1. A higher ACC stands for a better performance of corresponding

---

[1] http://archive.ics.uci.edu/ml/.

feature selection method. First of all, the figures shows that our proposed feature selection method achieved the best classification performance, followed by SOGFS, RUFS, CSFS and Baseline. For instance, our proposed method has improved by 1.31% and 6.31%, respectively, by comparing to the best comparison method (*i.e.*, SOGFS) and worst comparison method (*i.e.*, Baseline), on all four data sets.

Secondly, we can observe that the classification accuracy first increase to a peak and then gradually decreasing. For example, the ACC of our method increases from 85.19% when 10% feature preserved to 90.62% when 50% features preserved, and then fall to 87.50% when 90% feature preserved, in data set Ecoli. The reason can be supposed to that less feature can't provide enough useful information to train and obtain a robust model, while more features may introduce redundancy into the subset, which obviously corrupt the performance of classification. Besides, almost all feature selection methods outperform Baseline, it proves that the redundancy is common and could corrupt the performance of learning model.

Thirdly, by comparing to CSFS that separately considers about the global information, our method further takes local structure preservation into account, thus our method is able to output a better performance. By comparing to RUFS that considers about two structure preservation, our proposed method utilizes adaptive graph learning to yield a robust graph matrix and feature selection performance. Although SOGFS similarly use adaptive graph learning, our method further employed low-rank constraint to capture inherent structure of data from original data with redundancy. Thus, our proposed method achieved the best classification performance, which stands for our method can preserve the most informative features in the subset.

5.4   Parameters' Setting

There has two parameters need to tune in our model, that is α and γ. The variation of ACC with different parameter combination is listed in Fig. 2. We can observe that the ACC of our method slightly changed with the variation of parameter combination. For example, the ACC is about 93.1% in $\alpha = 10^{-3}, \gamma = 10^0$, while it is about 85.64% in $\alpha = 10^{-3}, \gamma = 10^{-1}$, at dataset *Ecoli*. Hence, it is easy to conduct a good classification performance via easily tuning parameters. As the limitation of pages, we only report two data sets, so as the convergence, and the remainder two data sets have similar situation.

Figure 3 reports the variation of the objective values (*i.e.*, Eq. (7)) with the increase of iteration. In our experiment, the stop criteria is set as $\frac{\|objv(t+1) - objv(t)\|_2^2}{objv(t)} \leq 10^{-3}$ for Algorithm 1, where $objv(t)$ indicates the objective function value of the t-th iteration of objective equation (*i.e.*, Eq. 7). From Fig. 3, it is obvious that our proposed optimization algorithm can fast converge, which usually within 20 iterations. Thus, our proposed optimization algorithm can effectively optimize the objective equation.

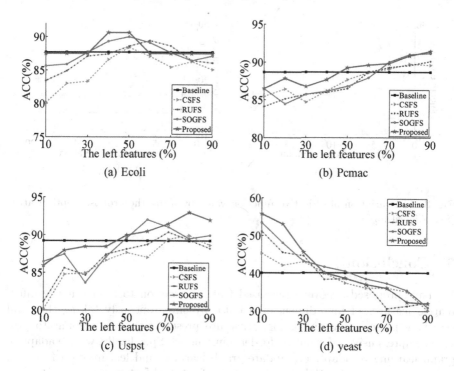

Fig. 1. The variation of ACC in different selected features.

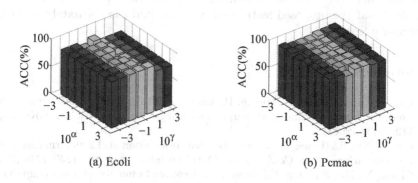

Fig. 2. The variation of ACC in different parameters' setting.

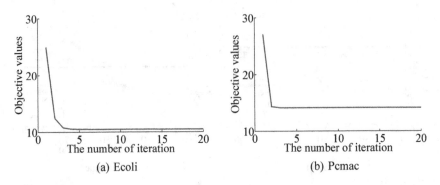

Fig. 3. The variation of objective function value by using the proposed optimization (*i.e.*, Algorithm 1).

6   Conclusion

We have proposed a novel supervised feature selection to filter out the informative features of high-dimensional data, by simultaneously taking the local structure preservation and global structure preservation into account to provide comprehensive information for learning model. Specifically, we use adaptive graph learning to adaptively update graph learning and feature data learning, so that it can output a global optimization of spectral feature selection. Besides, the low-rank constraint can help learning model capturing inherent structure so that can relieve the influence of redundancy. Extensive experimental results has exhibited that the proposed feature selection method can accurately select the informative features.

References

1. Cai, X., Ding, C., Nie, F., Huang, H.: On the equivalent of low-rank linear regressions and linear discriminant analysis based regressions. In: ACM SIGKDD, pp. 1124–1132 (2013)
2. Cai, X., Nie, F., Huang, H.: Exact top-k feature selection via l 2, 0-norm constraint. In: International Joint Conference on Artificial Intelligence, pp. 1240–1246 (2013)
3. Chang, X., Nie, F., Yang, Y., Huang, H.: A convex formulation for semi-supervised multi-label feature selection. In: AAAI, pp. 1171–1177 (2014)
4. Daubechies, I., DeVore, R.A., Fornasier, M., Güntürk, C.S.: Iteratively re-weighted least squares minimization: proof of faster than linear rate for sparse recovery. In: CISS, pp. 26–29 (2008)
5. Hu, R., et al.: Graph self-representation method for unsupervised feature selection. Neurocomputing **220**, 130–137 (2017)
6. Nie, F., Zhu, W., Li, X.: Unsupervised feature selection with structured graph optimization. In: AAAI, pp. 1302–1308 (2016)
7. Nie, F., Zhu, W., Li, X.: Unsupervised feature selection with structured graph optimization. In: Thirtieth AAAI Conference on Artificial Intelligence, pp. 1302–1308 (2016)

8. Qian, M., Zhai, C.: Robust unsupervised feature selection. In: IJCAI, pp. 1621–1627 (2013)

9. Li, Y., Zhang, J., Yang, L., Zhu, X., Zhang, S., Fang, Y.: Low-rank sparse subspace for spectral clustering. IEEE Trans. Knowl. Data Eng. https://doi.org/10.1109/TKDE.2018.2858782

10. Zhang, S., Li, X., Zong, M., Zhu, X., Wang, R.: Efficient kNN classification with different numbers of nearest neighbors. IEEE Trans. Neural Netw. Learn. Syst. **29**(5), 1774–1785 (2018)

11. Zheng, W., Zhu, X., Wen, G., Zhu, Y., Yu, H., Gan, J.: Unsupervised feature selection by self-paced learning regularization. Pattern Recogn. Lett. (2018). https://doi.org/10.1016/j.patrec.2018.06.029

12. Zheng, W., Zhu, X., Zhu, Y., Hu, R., Lei, C.: Dynamic graph learning for spectral feature selection. Multimed. Tools Appl. (2017). https://doi.org/10.1007/s11042-017-5272-y

13. Zhu, P., Zuo, W., Zhang, L., Hu, Q., Shiu, S.C.K.: Unsupervised feature selection by regularized self-representation. Pattern Recogn. **48**(2), 438–446 (2015)

14. Zhu, X., Huang, Z., Yang, Y., Shen, H.T., Xu, C., Luo, J.: Self-taught dimensionality reduction on the high-dimensional small-sized data. Pattern Recogn. **46**(1), 215–229 (2013)

15. Zhu, X., Li, X., Zhang, S., Ju, C., Wu, X.: Robust joint graph sparse coding for unsupervised spectral feature selection. IEEE Trans. Neural Netw. Learn. Syst. **28**(6), 1263–1275 (2017)

16. Zhu, X., Li, X., Zhang, S., Xu, Z., Yu, L., Wang, C.: Graph PCA hashing for similarity search. IEEE Trans. Multimed. **19**(9), 2033–2044 (2017)

17. Zhu, X., Suk, H.-I., Huang, H., Shen, D.: Low-rank graph-regularized structured sparse regression for identifying genetic biomarkers. IEEE Trans. Big Data **3**(4), 405–414 (2017)

18. Zhu, X., Wu, X., Ding, W., Zhang, S.: Feature selection by joint graph sparse coding (2013)

19. Zhu, X., Zhang, S., Hu, R., Zhu, Y., et al.: Local and global structure preservation for robust unsupervised spectral feature selection. IEEE Trans. Knowl. Data Eng. **30**(3), 517–529

20. Zhu, Y., Kim, M., Zhu, X., Yan, J., Kaufer, D., Wu, G.: Personalized Diagnosis for Alzheimer's Disease. In: Descoteaux, M., Maier-Hein, L., Franz, A., Jannin, P., Collins, D.L., Duchesne, S. (eds.) MICCAI 2017. LNCS, vol. 10435, pp. 205–213. Springer, Cham (2017). https://doi.org/10.1007/978-3-319-66179-7_24

21. Zhu, Y., Lucey, S.: Convolutional sparse coding for trajectory reconstruction. IEEE Trans. Pattern Anal. Mach. Intell. **37**(3), 529–540 (2015)

22. Zhu, Y., Zhu, X., Kim, M., Kaufer, D., Wu, G.: A novel dynamic hyper-graph inference framework for computer assisted diagnosis of neuro-diseases. In: Niethammer, M., et al. (eds.) IPMI 2017. LNCS, vol. 10265, pp. 158–169. Springer, Cham (2017). https://doi.org/10.1007/978-3-319-59050-9_13

23. Zhu, Y., Zhang, X., Hu, R., Wen, G.: Adaptive structure learning for low-rank supervised feature selection. Pattern Recogn. Lett. **109**, 89–96 (2018)

24. Zhu, Y., Zhong, Z., Cao, W., Cheng, D.: Graph feature selection for dementia diagnosis. Neurocomputing **195**(C), 19–22 (2016)

A Tree Based Approach for Concept Maps Construction in Adaptive Learning Systems

Niharika Gupta, Vipul Mayank, M. Geetha⬤, Shwetha Rai(✉)⬤, and Shyam Karanth⬤

Department of Computer and Engineering, Manipal Institute of Technology, Manipal Academy of Higher Education, Manipal 576104, Karnataka, India
niharika3gupta@gmail.com, vipulmayank93@gmail.com, {geetha.maiya, shwetha.rai, shyam.karanth}@manipal.edu

Abstract. A concept map is a diagram that depicts suggested relationships between concepts. The relationships are marked by a relevance degree that denotes the level of correlation between any two concepts. Concept map is a graphical tool used to structure and organize knowledge. In this project, a concept map will be generated based on a real-life dataset of how questions are answered by students (correctly or incorrectly) and the weight of the concepts in the questions. Several algorithms have been proposed to automatically construct concept maps. However, all these algorithms use Apriori algorithm to discover the frequent itemsets and get the association rules. Apriori algorithm requires several database scans, and thus, it is not efficient. A tree-based approach (i.e., FP tree algorithm) adopted in this project to overcome the drawbacks of the Apriori algorithm in the construction of concept maps for adaptive learning systems.

Keywords: Apriori · Association rules · Concept map · Frequent itemset Tree based

1 Introduction

Data mining is a computational process of discovering interesting patterns in an enormous amount of data sets. It involves methods at the intersection of different areas such as artificial intelligence, database systems, machine learning, and statistics. The knowledge extracted by using data mining techniques should be novel and useful.

For centuries the extraction of patterns from data was done manually. Over the period as datasets have increased, the complexity has grown. The manual data analysis has increasingly improved with automated data processing, supported by other techniques such as cluster analysis, decision trees and decision rules, genetic algorithms, neural networks, etc.

Data mining is prominently used in Educational mining field. Many researchers [1–12] have worked in this area to analyze the student performance, which aided in syllabus preparation for the teachers, give more prominence to the students who do not understand specific topics, etc.

© Springer Nature Singapore Pte Ltd. 2018
Q. Chen et al. (Eds.): ATIS 2018, CCIS 950, pp. 172–182, 2018.
https://doi.org/10.1007/978-981-13-2907-4_15

Concept maps are a diagrammatic representation used to represent the relationship between the concepts. It has been developed to improve the quality of meaningful learning in the sciences. Concept mapping aids in various purposes for learners such as:

- Encouraging students to generate new ideas.
- Allowing students to discover and evaluate new concepts and the plans that relate them to previously learned concepts.
- Helping students to communicate their ideas and thoughts more evidently.
- Enabling students to integrate old and newer concepts.
- Helping students to gain quality knowledge of any topic and process the information.

A lot of focus is given to research topics based on adaptive learning systems in recent years. Several methods have been proposed to automatically construct concept maps. This method will enable students to learn more meaningfully. Precise concept maps will be created by applying the Association Rule data mining technique.

In this paper, the proposed method implements a tree-based approach that will be more time and space efficient than the currently existing algorithms.

2 Literature Review

Several algorithms have been proposed over the past years to generate concept maps for adaptive learning systems automatically. Most of the algorithms are variations of one another wherein they try to eliminate the drawbacks of the algorithm on which they are based. They all generally follow the same vein of thought wherein the data mining technique of association rule mining is employed. Four algorithms are referred to in this project. All these algorithms use Apriori algorithm for the generation of frequent itemsets and subsequently, association rules.

Lee et al. [8] had proposed an approach in which they examined students' assessment data in binary grades by applying a data mining method. The technique used was the association rules to construct concept maps. In the approach, they generated first generated all the frequent itemsets using Apriori algorithm. Next, the questions which had been answered incorrectly by the students were considered. Based on the frequent itemsets, they generated association rules and found the relevance degree between any two concepts. These relevance degrees were used in the final construction of the concept maps.

Chen and Bai [9] presented a technique to construct concept maps automatically for adaptive learning systems using data mining methods. They sought to improve upon Lee et al.'s technique wherein they overcame its drawback of only considering failure to failure association rules. They considered failure-to-failure association rules (according to Lee et al.'s method) and considered correct-to-correct association rules. This approach was found to give a more accurate concept map than the previously explained algorithm. However, even this method may construct incorrect concept maps in some situations.

Chen and Sue [10] proposed a new algorithm for automatically constructing concepts maps based on data mining techniques. Initially, the value that indicates the answer-consistence between given two questions is calculated. This value is known as Counter Value. Next, four types of association rules are considered between two questions to extract relevant information using the previously calculated counter values. The four types of association rules are:

- Failure-to-failure relationship
- Correct-to-correct relationship
- Correct-to-failure relationship
- Failure-to-correct relationship.

Finally, the relevance degree is calculated from the derived association rules for any two concepts. This method overcomes the disadvantages of Chen and Bai's method [9]. This algorithm was found to be the most accurate among the three methods discussed so far for the construction of concept maps.

Most proposed methods only consider binary grades of each test item i.e. they only take into consideration whether the question has been answered correctly or not. However, in a real assessment environment, marks are generally assigned to each question and how much a learner score out of the maximum allotted marks for a question may also be used to assess the learner's level of understanding. Furthermore, the existing methods that are based on fuzzy set theory do not analyze the weight of concept in each question, and that may lead to the construction of incorrect relationships or amplify the degree of relationship. Al-Sarem et al. [11] proposed an approach to construct concept maps automatically by analyzing the result of analysis of numerical testing scores after applying fuzzy set theory.

Shieh and Yang [12] proposed a method to improve the accuracy of the outcome since some of the student data records may not follow a normal distribution and many algorithms proposed methods based on the normal distribution. A Student-Problem Chart technique was applied to six different performance groups.

3 Problem Definition

One of the most important research topics of learning systems is the construction of concept maps that provide learning guidance to learners. The drawbacks of the existing algorithms are as follows:

- Most of the algorithms only considered either the correctly answered questions or the incorrectly answered questions.
- The existing algorithms use the Apriori algorithm for the generation of frequent itemsets and the association rules. However, the Apriori algorithm requires multiple database scans to generate the frequent itemsets. Hence it is not time and space efficient.

The drawbacks are overcome as follows:

- Questions answered both correctly and incorrectly by learners are considered to glean useful information properly, otherwise, some relationships between concepts will be lost, and unnecessary relationships may be built in the constructed concept maps.
- FP tree algorithm will be implemented which requires only two database scans to generate all frequent itemsets.

4 Methodology

The flowchart that is given below (see Fig. 1) depicts the entire process of finding the relevance degree which in turn will lead to the creation of concept maps.

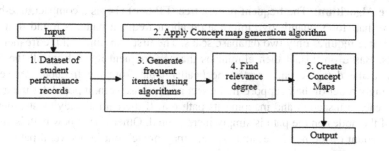

Fig. 1. Flowchart of the entire process of calculating relevance degree

The elements in the flowchart are detailed as follows:

1. The real-life dataset used in this project consists of the marks obtained by n learners in m questions and the weight of p concepts in those questions. In this project, we have taken $n = 60$ and $m = 3$.
2. In this paper, the Chen and Bai [9] approach have been implemented.
3. The frequent itemsets are initially generated using Apriori in the existing algorithms. However, this method is not efficient for large databases. Hence, the FP algorithm has been implemented for the generation of frequent itemsets since this algorithm has lesser space and time complexity than Apriori.
4. The relevance degree determines the degree of dependence of a concept on another in a question.
5. Finally, based on the relevance degrees, a concept map is constructed where the edges between any two concepts are the maximum of all the relevance degrees between them.

4.1 Calculation of Frequent Itemsets

The frequent itemsets have been generated using two algorithms namely Apriori and Frequent Pattern (FP) tree algorithm.

Apriori Algorithm: Apriori [13] is an algorithm for discovering frequent itemsets and association rule learning over a given transactional database. It begins by discovering the frequent one itemsets in the given database and then extends them to larger itemsets if those itemsets appear often enough in the database. Support denotes the total number of occurrences of a pattern in the database. If the support of an itemset is greater than the user-defined support, i.e., minimum support, then the itemset is said to be frequent. The frequent itemsets determined by Apriori can be used to identify association rules, and these emphasize the general trends in the database. This technique is initially applied in the market basket analysis. The main drawback of the Apriori is that it may require multiple scans of the database and hence it is computationally expensive, especially when there are a lot of patterns or if the pattern length is especially long.

FP Tree Algorithm: The frequent-pattern tree (FP-tree) [14] is a compact, tree-based structure that stores quantitative information about frequent patterns found in a database. FP tree requires only two database scans. The first scan finds all the frequent one itemsets. Based on this, it then eliminates the non-frequent items from all the transactions in the database and creates the header table from the frequent itemsets in decreasingly order of their support. In the second database scan, it populates the tree by reading each transaction and mapping its path out. If the path already exists, then the count of the nodes on the path is simply incremented. Otherwise, a new path is created with the count of the nodes set to 1. A path may branch out into several paths.

4.2 Finding the Relevance Degree

Association rules that help discover relationships between the data in a transactional database. Association rules are created by analyzing data for frequent if/then patterns and using the conditions of user-defined minimum support and user-defined minimum confidence to identify the relationships that are of more interestingness. The confidence indicates the total number of times the interesting patterns have been found to be present in the transactional database.

Once the frequent itemsets have been found, the association rules are generated from them which are used in finding relevance degree using the Eq. 1.

$$\text{rev}\left(C_i \rightarrow C_j\right)_{Q_x Q_y} = \text{Min}\left(W_{Q_x C_i}, W_{Q_y C_j}\right) \times \text{conf}\left(Q_x \rightarrow Q_y\right) \tag{1}$$

Where,

- $\text{rev}\left(C_i \rightarrow C_j\right)_{Q_x Q_y}$ denotes the relevance degree of the relationship "$C_i \rightarrow C_j$" converted from the relationship "$Q_x \rightarrow Q_y$", $\text{rev}\left(C_i \rightarrow C_j\right)_{Q_x Q_y} \in [0, 1]$.
- C_i represents a concept that appears in the question Q_x, and C_j represents a concept that appears in the question Q_y.

- $W_{Q_x C_i}$ represents the weight associated with the concept C_i in the question Q_x, and $W_{Q_y C_j}$ represents the weight associated with the concept C_j in the question Q_y.
- $1 \leq i \leq tc$, $1 \leq j \leq tc$, $1 \leq x \leq tq$, $1 \leq y \leq tq$, where tc is the total number of concepts and tq is the total number of questions.

If there are multiple relationships between any two concepts, then the relationship with maximum relevance degree is selected to denote the degree of the relationship between the two concepts. Thus, a concept relationship table is created.

4.3 Constructing Concept Map

Once the concept relationship table has been generated, the concept map can be constructed. A concept map consists of nodes and links. The nodes represent the concepts and the links i.e., edges between the nodes represent the relevance degree between the concepts. To construct a concept map, for each relationship $C_i \rightarrow C_j$ in the concept relationship table, an edge is added from concept C_i to concept C_j in the concept map with the relevance degree of relationship $C_i \rightarrow C_j$. An example of a concept map is given below (see Fig. 2).

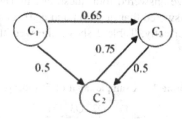

Fig. 2. Diagrammatic representation of a concept map

4.4 Methodology of the Improved Algorithm

The flowchart of the improved algorithm is depicted below (see Fig. 3).

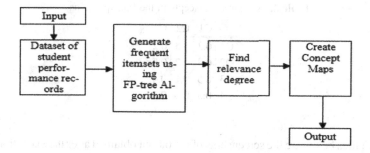

Fig. 3. Methodology of the improved algorithm

The modules are as follows:

1. The input is the real-life dataset of 60 students and four questions. The dataset details in binary grade whether the students have answered the questions correctly or not.
2. The FP Tree algorithm has been coded in Java, and it outputs the frequent itemsets in a text file.
3. Frequent Itemsets refers to the output file generated by running FP tree construction code. It is the input to the next module.
4. Find relevance degrees and Create concept map modules have been coded in R language.
5. The constructed concept map is the output.

5 Results and Analysis

5.1 Results

The input taken for the implementation of the two algorithms is a real-life dataset. In the dataset, 60 students have answered four questions. In Table 1 A "1" denotes that the student has correctly answered that question and a "0" denotes that the student has answered the question incorrectly. Table 2 shows the weight of concepts in the four questions.

Table 1. A sample input of 5 students

	Q1	Q2	Q3	Q4
S1	1	1	1	0
S2	1	1	0	0
S3	0	0	0	0
S4	1	1	0	0
S5	1	1	1	0

Table 2. Weight of concepts in the four questions

	C1	C2	C3
Q1	0.3	0.5	0.2
Q2	0.3	0.2	0.5
Q3	0.5	0.1	0.4
Q4	0.3	0.4	0.3

The figures below are the screenshots of the output obtained after the execution of the code. The first figure (see Fig. 4) contains all the generated frequent itemsets. The second figure (see Fig. 5) contains all the generated frequent itemsets along with their relevance degrees which in turn would be required for the construction of the concept maps.

[1] "Sessional 1: No. of students who answered the questions correctly\n"
      Q1   Q2   Q3   Q4
      [,1] [,2] [,3] [,4]
[1,]   44   38   10   2
[1] "Sessional 1: No. of students who answered the questions incorrectly\n"
      [,1] [,2] [,3] [,4]
[1,]   16   22   50   53
[1] "Sessional 2: No. of students who answered the questions correctly\n"
      [,1] [,2] [,3] [,4]
[1,]   41   7   32   35
[1] "Sessional 2: No. of students who answered the questions incorrectly\n"
      [,1] [,2] [,3] [,4]
[1,]   19   53   28   24
Frequent 1 itemsets for correctly answered questions in sessional 1 are :
[1] 1 2
Frequent 1 itemsets for incorrectly answered questions in sessional 1 are :
[1] 3 4
Frequent 1 itemsets for correctly answered questions in sessional 2 are :
[1] 1 3 4
Frequent 1 itemsets for incorrectly answered questions in sessional 2 are :
[1] 2 3 4

Fig. 4. Screenshot of the generated frequent itemsets

[1] 1 2

Its relevance degree is
[1] 0.075
[[1]]
[1] 1 3

Its relevance degree is
[1] 0.075
[[1]]
[1] 1 4

Its relevance degree is
[1] 0.045
[[1]]
[1] 2 3

Its relevance degree is
[1] 0.08533333
[[1]]
[1] 2 4

Its relevance degree is
[1] 0.02666667
[[1]]
[1] 3 4

Its relevance degree is
[1] 0.0306923
Time difference of 0.332567 secs

Fig. 5. Screenshot of the final output

5.2   Analysis

In this paper, a comparative review of 2 frequent itemsets generating algorithms, namely, Apriori and Frequent Pattern tree, has been done.

Apriori algorithm is the most basic algorithm for the generation of frequent itemsets. It follows a bottom-up approach in which the frequent itemsets are extended to one item at a time. Thus, it is a very slow algorithm as it requires multiple database scans.

FP tree generates all the frequent itemsets in just two database scans hence making it a much faster and more efficient algorithm that Apriori. A comparison between the execution times of these two algorithms has been done. Ten execution times have been noted for both the algorithms and plotted in a line graph.

Furthermore, the average execution times have been calculated for both the algorithms and plotted in a bar graph.

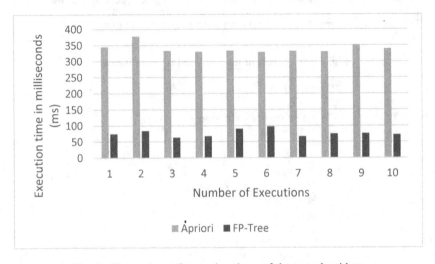

Fig. 6. Comparison of execution times of the two algorithms

Hence by the above bar graph (see Fig. 6), it can be concluded that FP tree is, evidently, more time efficient than Apriori algorithm. FP tree uses a hash-map structure to store quantitative information about the frequent itemsets found in the database. Apriori uses array structures to store the generated frequent itemsets. An associative array is implemented using a hash-map data structure which enables to assign keys to values. Hence a hash-map is a concise representation of an array.

Table 3. Time complexity of a hash map

Parameter	Average case	Worst case
Space	O(n)	O(n)
Search	O(1)	O(n)
Insert	O(1)	O(n)
Delete	O(1)	O(n)

Tables 3 and 4 shows the comparison of the time and space complexities of the hash-map and array structure.

Table 4. Time and space complexity of array structure

Parameter	Average case	Worst case
Space	O(n)	O(n)
Search	O(n)	O(n)
Insert	O(n)	O(n)
Delete	O(n)	O(n)

Thus, it can be concluded that hash-map is the superior data structure thus making the FP tree algorithm the more time and space efficient algorithm over the Apriori algorithm.

6 Conclusion

Most of the concept map construction algorithms followed Apriori approach to discover the frequent patterns and construction a concept map. Since the time taken by the Apriori algorithm to discover frequent patterns increases enormously with time, a non-candidate generation algorithm, FP-tree, was implemented. Once the frequent itemsets were generated by using FP Tree algorithm, the relevance degrees were calculated for the various concept-concept relationships. Generated the concept map based on the provided data of the weightage of concepts in questions and using the real-life dataset of detailing whether students have answered the questions correctly. The results of performance analysis corroborated the premise that Apriori algorithm is less efficient algorithm compared to FP Tree.

This work can be improved in several ways. FP-tree algorithm takes two scans to discover frequent itemsets. This algorithm can be improved by implementing algorithms which use a single database scan to discover the frequent itemsets. Only binary grade data of students has been considered. However, most of the time numeric marks are assigned to answers. Hence numerical analysis may be conducted.

References

1. Romero, C., Ventura, S.: Educational data mining: a review of the state of the art. IEEE Trans. Syst. Man Cybern. Part C: Appl. Rev. **40**(6), 601–618 (2010)
2. Yoo, J.S., Cho, M.H.: Mining concept maps to understand university students' learning. In: Yacef, K., Zaïane, O., Hershkovitz, A., Yudelson M., Stamper, J. (eds.) Proceedings of the 5th International Conference on Educational Data Mining, Chania, Greece, vol. 41, pp. 184–187 (2012)

3. Kavitha, R., Vijaya, A., Saraswathi, D.: An augmented prerequisite concept relation map design to improve adaptivity in e-learning. In: Proceedings of the International Conference on Pattern Recognition, Informatics and Medical Engineering, pp. 8–13. IEEE, Salem (2012)

4. Kardan, A., Imani, M.B., Ebrahim, M.A.: ACO-map: a novel adaptive learning path method. In: Proceedings of the 4th International Conference on E-Learning and E-Teaching, pp. 20–25. IEEE, Shiraz (2013)

5. Acharya, A., Sinha, D.: Construction of automated concept map of learning using hashing technique. In: Satapathy, S.C., Biswal, B.N., Udgata, S.K., Mandal, J.K. (eds.) Proceedings of the 3rd International Conference on Frontiers of Intelligent Computing: Theory and Applications (FICTA) 2014. AISC, vol. 327, pp. 567–578. Springer, Cham (2015). https://doi.org/10.1007/978-3-319-11933-5_64

6. Acharya, A., Sinha, D.: A weighted concept map approach to generate learning guidance in science courses. In: Mandal, J.K., Satapathy, S.C., Sanyal, M.K., Sarkar, P.P., Mukhopadhyay, A. (eds.) Information Systems Design and Intelligent Applications. AISC, vol. 340, pp. 143–152. Springer, New Delhi (2015). https://doi.org/10.1007/978-81-322-2247-7_16

7. Acharya, A., Sinha, D.: An intelligent web-based system for diagnosing student learning problems using concept maps. J. Educ. Comput. Res. 55(3), 1–23 (2016)

8. Lee, C., Lee, G., Leu, Y.: Application of automatically constructed concept map of learning to conceptual diagnosis of e-learning. Expert Syst. Appl.: Int. J. 36(2), 1675–1684 (2009)

9. Chen, S., Bai, S.: Using data mining techniques to automatically construct concept maps for adaptive learning systems. Expert Syst. Appl.: Int. J. 37(6), 4496–4503 (2010)

10. Chen, S., Sue, P.: Constructing concept maps for adaptive learning systems based on data mining techniques. Expert. Syst. Appl.: Int. J. 40(7), 2746–2755 (2013)

11. Al-Sarem, M., Bellafkih, M., Ramdani, M.: Mining concepts' relationship based on numeric grades. Int. J. Comput. Sci. 8(3), 136–142 (2011)

12. Shieh, J.C., Yang, Y.T.: Concept maps construction based on student-problem chart. In: Proceedings of IIAI 3rd International Conference on Advanced Applied Informatics, pp. 324–327. IEEE, Kitakyushu (2014)

13. Agarwal, R., Srikant, R.: Fast algorithm for mining association rules. In: Bocca, J.B., Jarke, M., Zaniolo, C. (eds.) Proceedings of the 20th International Conference on Very Large Data Bases, pp. 487–499. ACM, Morgan Kaufmann Publishers Inc., San Francisco (1994)

14. Han, J., Pei, J., Yin, Y.: Mining frequent patterns without candidate generation. In: Chen, W., Naughton, J., Bernstein, P.A. (eds.), Proceedings of ACM SIGMOD International Conference on Management of Data, pp. 1–12. ACM, Dallas (2000)

Applications

Active Multivariate Matrix Completion

Tianshi Liu$^{(\boxtimes)}$, Qiong Wu, Wenjuan Zhang, and Xinran Cao

Xi'an Shiyou University, Xi'an, China

`232228949@qq.com, qwu@tulip.academy, 137956371@qq.com, 446225268@qq.com`

Abstract. Matrix completion has been rapidly grown interested in areas of engineering and applied science. however, most of applications are aim at single variate matrix. In fact, majority applications datasets express in form of multivariate matrix, developing multivariate matrix application is very necessary. The important is when the missing values too many, matrix completion of take advantage of active learning performance better than standard matrix completion. Although, several combining active learning for matrix completion solutions have been proposed, and most of them based on query strategy. But none of them emphasize the important location of matrix for recovering a matrix. In this paper, we design *active multivariate matrix completion*. The goal of this algorithm is that find a important location of matrix for matrix completion and combine rank aggregation to select query entries. Experiment evaluation base on images datasets. when we query a small amount of missing entries, the proposed *active multivariate matrix completion* efficiently raise the accuracy of matrix completion and give the important position of missing entries.

Keywords: Matrix completion · Active learning · Rank aggregation

1 Introduction

A *Wireless sensor network* (WSN) could cooperatively transfer their sensed data to central servers, thus the spatially distributed sensor nodes can be monitor the environment [1]. Since WSNs consists of many sensor nodes, the key research issue in WSNs is reducing energy consumption to prolong the practical lifetime. To address this issue, the most effective way is to reduce the number of collected data, which causes the data recovery problem to recover the sensor nodes data in WSNs.

A common approach for data recovery problem is *matrix completion*. In general, accurately recovery a matrix from a small number of known entries is impossible, but if the unknown matrix is low rank that can be radically changes this premise, making the search for solutions meaningful [2]. However, in many cases, the observed entries proportion is very small. *Candés* and *Recht* proved that an $n_1 \times n_2$ matrix of rank r with large probability to succeed recovery matrix, the number of observed entries obeys $m \geq C n^{\frac{6}{5}} r \log n$. Therefore, there has

© Springer Nature Singapore Pte Ltd. 2018
Q. Chen et al. (Eds.): ATIS 2018, CCIS 950, pp. 185–198, 2018.
https://doi.org/10.1007/978-981-13-2907-4_16

been tremendous development of design active matrix completion to leverage human intelligence to improve the probability to succeed recovery matrix and performance of matrix completion. Most of active matrix completion algorithms combine query strategy to pick some entries for human annotation, with the aim of minimizing the reconstructive error of the estimated matrix and find the important position entries to reconstruct matrix. Along with the development of matrix completion, it has become an important mathematics model to constructing a complete data matrix from an incomplete data matrix for recommendation and prediction.

Matrix completion has been widely applied in many applications, such as recommendation and prediction research community. However, majority existing work only focus on single variate problem. Despite this, real-world datasets are expressed in the form of multivariate. Such occasions include WSNs systems (where entries of matrix indicated the node's status), computer vision (RGB image matrices). While matrix completion based on WSNs system is classic application, the problem is majority of WSNs system are multivariate. We refer to this situation is multivariate matrix completion. Hence, the most important to resolve is how to integrate single variate matrix completion to multivariate matrix completion as soon as possible.

We design an active approach for the multivariate matrix completion, this solution incorporating the rank aggregation to the single variate matrix completion can which help tackling the research barrier. The challenges of this solution detail as below:

- The first challenges is how to integrate single variate matrix completion to multivariate matrix completion. Most of the applications the data express as multivariate, exploration of the multivariate matrix is very important to explore which variable is more important and how they are interacting with others.
- The second challenges is how to integrate rank aggregation with multivariate matrix completion. Rank aggregation can help us to decide which area of matrix is most important and how they are influence recommendation and prediction.

The above challenges imply that the matrix completion unable incorporated into multivariate and rank aggregation with a straight forward measure. There is a high demand to design a novel mechanism combining them. For the first challenge, many applications datasets are expressed as multivariate matrix, but more or less they have correlated with single variate matrix. In other words, we can transforming the multivariate matrix into several univariate matrixs. Based on this, we proposed the multivariate matrix completion. The second challenge can be solved by combining active learning with rank aggregation. If some queries can be answered by the biggest score, the queries mechanism will be very persuasiveness. Based on above observations, we present the solution to active multivariate matrix completion, the contributions of this paper can be summarized as below:

- First, We theoretically analyse the applications and challenges of proposed active multivariate matrix completion. Base on this, presenting a clear problem definition and exploring the significance of active multivariate matrix completion for recommendation and prediction
- Second, we design an active learning mechanism to actively select a number of entries for human annotation.
  This solution select entries by the entropy values of rank aggregation. Thus we demonstrate our select mechanism is more authority than others. The major advantage of this solution is decreasing noise for each query.
- Third, we propose the notion of important location of matrix based on the active learning with rank aggregation, which helps decrease the prediction cost for a missing values multivariate matrix.

The rest of this paper is organized as follows. Section 2 We present the preliminaries and related work. Section 3 discusses rank aggregation with active learning, the active multivariate matrix completion theoretical explanation and utility analysis of the active multivariate matrix completion. Section 4 presents experimental results, followed by conclusions in Sect. 5.

2 Preliminaries and Related Work

This section reviews two fundamental concepts: *Matrix Completion with Queries* and *Rank Aggragation*, then briefly surveys the related works in *Matrix Completion* and *Active Matrix Completion*.

2.1 Preliminaries

Matrix Completion with Queries. In practice, the queries strategy is a common strategy for active learning. Picks some entries from missing entries which have attribute deserving manual annotation. Evaluation criterion are very different from each other.

Given a multi-variate normal distribution variable X (of dimensionality p). naturally, we get mean vector μ and covariance matrix Σ. Thus, $X \sim \mathcal{N}(\mu, \Sigma)$. For example, a random sample $x = (x_1, x_2, ..., x_n)^T$, then parameter μ and Σ can be computed. In this paper, the concentration matrix denote as J, $J = \Sigma^{-1}$. Minimize the negative l_1 penalized the typical approach is log-likelihood [3]:

$$- L(\mu, J; x) + \lambda \|J\|_1 = -\frac{n}{2}|J| + \frac{1}{2}\sum_{i=1}^{n}(x_i - \mu)^T J(x_i - \mu) + \lambda \|J\|_1 \quad (1)$$

Base on their uncertainty (variance) values sort the missing entries in descending order. The top k uncertain entries can query the ground truth values from manual annotation.

Rank Aggregation. Consider S is a set of k objects, thus a sequenced list of size k corresponding the objects of S. When the ordered list $l_1, l_2, \ldots, l_n$ are given, the problem of rank aggregation formulation as below:

$$l^* = \underbrace{\arg min}_{l_c} \sum_{i=1}^{n} w_i \times d(l_c, l_i) \qquad (2)$$

where l^* represent the consensus ordering of the objects in S, thus represent minimum value of the cumulative disagreement with the given ordering is minimum. l_c denotes a candidate for l^*, w_i denotes the weight assigned to l_i. In the current work, w_i differently takes the values of unity, d represents a certain of the disagreement between two ordering (or list).

2.2   Related Work

Statistical Matrix Completion. The statistical matrix completion attract scientist tremendous attention since 2000s' [4]. and many of them have been extended recommender system [5] and Wireless Sensor Networks [6]. Statistical approaches typically with the assumption of the observed matrix is low-rank structure. Besides, those methods usually assume the positions of known entries follow a random distribution, thus the matrix completion task was treated as an optimization formulation. The key characteristic of statistical method is that regardless of the observed entries is sufficient. That is to say, there must have best estimate of output for any input, which cause high error and high time complexity. Moreover, when the information is quite insufficient, statistical methods are invalid for completion mechanism any more.

    Candés and *Recht* presented if the observed entries amounts too few, it's hard to accomplish a high accuracy of matrix completion. Thus they proposed a threshold value to guarantee successful do matrix completion [4]. They demonstrate that an $n_1 \times n_2$ matrix of rank r should have at lest $m > C n^{\frac{6}{5}} r \log(n)$ observed entries to reach a successful matrix completion. where $n = max(n_1, n_2)$. However, satisfy this bound for a real word datasets must have a large amount of observed entries. For example, the famous solutions of Netflix Challenge adopt the rank $r \approx 40$, there must have over 151 million observed entries to guarantee high accuracy matrix completion.

Active Matrix Completion. Active learning has been developed for many aspects, the representativeness samples of unlabeled data are automatically found out to train a classifier. This measure in certain of tremendous reduce the cost of human annotation and raise the classifier accuracy. Base on information content, there has been studied querying strategy with example for matrix completion in the context of adaptive sampling in [7]. Such problem has been used in many specific applications, for example, matrix factorization in some aspects improve through querying values with a bayesian method [8], spectral clustering also has been improved through querying pairwise similarities [9] or even application of semi-supervised link prediction through the context information [10].

Those works mainly focus on deciding which entries need to be annotation aim at reduce error in the matrix reconstruction process. Reveal entries evaluation criterion are very different from each other. For example, some missing entries have the largest uncertainty or the biggest score. In [11] the authors assumed for each case (each row or column), the set of missing entries and observed entries satisfy conditional multivariate normal distribution. Naturally, we obtain the missing entries mean vector and covariance matrix. Additionally, Each diagonal elements values (uncertainty) of covariance matrix corresponding with each missing entries. And the top k diagonal elements values entries will be selected for human annotation. In [12] focus on finding a promising ordering with input matrix. The ordering of queries affects the accuracy of the completion, base on this combining linear system can be only asking a small number of queries. In [13] the mask graph each nodes corresponds to the matrix entries, they assumed the rows (or columns) of matrix entries relationship associated with the observed entries as indicators for the edges. The entries can be queried which guarantee the mask graph is chordal after these queries and has the higher informative.

Summary. As a result of the lack exploitation of multivariate matrix backgrounds, existing research focus on complete matrix completion fails to provide practical solutions to combine the multivariate and active matrix completion. However, with the limitation of existing works in active matrix completion, the purpose of this paper is filling this void through incorporating multivariate matrix with an active matrix completion that aim to find the important position for completion and high quality of matrix completion. More particularly, we have to face the following issues:

– How to cooperate the matrix completion with multivariable matrix ?
– How to identify the important location of matrix for matrix completion ?

3   Active Multivariable Matrix Completion

In this section, for developing single variate matrix completion to multivariate matrix completion which can suit more situation, we first propose Active Multivariate Matrix Completion.

3.1   Selection Entries from Incomplete Matrix

We assumed a particular case which form of each row (or column) of the data matrix.

For each case, in mathematical we supposed that each case the set of missing entries satisfy multivariate normal distribution. Thus matching the mean vector and covariance matrix. The covariance matrix of missing entries each diagonal elements associated with each missing entries. The overarching idea is construct a rank list which from missing entries associated with covariance matrix diagonal elements. The high values of diagonal elements associated with missing entries was selected. kemeny optimal aggregation to evaluate whether the entries is important, and the top k values of missing entries with manual annotation. We design this algorithm with the foundation of GLasso and Miss-Glasso frameworks [14]. The frameworks was raised to solve the sparse inverse covariance estimation, for each case missing entries follows multivariate normal distribution. The details of these methods was presented as follows.

MissGlasso: Stadler and Buhlman proposed MissGlasso focus on estimating the mean and covariance matrix of missing entries. When the multivariate normal distribution of missing entries condition on observed entries [14]. Here the mean vector μ and covariance matrix Σ of $(X^{(1)}, \ldots, X^{(p)}) \sim \mathcal{N}(\mu, \Sigma)$ is P-variate normally distributed parameter. The observed entries and missing entries was denoted as $x = (x_{obs}, x_{mis})$, Then the size n of random sample can be expressed as

$$X_{obs} = (x_{obs,1}, x_{obx,2}, \ldots, x_{obs,n}, n)$$

$x_{obs,i}$ indicates the observed entries in case $i, i = 1, \ldots, n$. For each case the missing entries covariance matrix can be well estimate from likelihood function:

$$l(\mu, \Sigma; x_{obs}) = -\frac{1}{2} \sum_{i=1}^{n} (\log |\sum_{obs,i}| + (x_{obs,i} - \mu_{obs,i})^T (\sum_{obs,i})^{-1} (x_{obx,i} - \mu_{obs,i})) \quad (3)$$

$\mu_{obs,i}$ and $\sum_{obs,i}$ denote the mean vector and covariance matrix of the observed entries. Equivalently, express as J:

$$l(\mu, \Sigma; x_{obs}) = -\frac{1}{2} \sum_{i=1}^{n} (\log |(J^{-1})_{obs,i}| + (x_{obs,i} - \mu_{obs,i})^T ((J^{-1})_{obs,i})^{-1} (x_{obx,i} - \mu_{obs,i}))$$
$$(4)$$

Over all the cases, μ and J can be estimated from the observed log-likelihood. Combine the L_1 penalty with the concentration matrix J:

$$\hat{\mu}, \hat{J} = \underset{(\mu,J):J \succ 0}{\arg \min} -L(\mu, J; x_{obs}) + \lambda \|J\|_1 \quad (5)$$

The above problem can be solved by partitioned Gaussians. Suppose a partition $(X_1, X_2) \sim \mathcal{N}(\mu, \Sigma)$). Thus the mean $\mu_2 + \Sigma_{21}\Sigma_{11}^{-1}(X_1 - \mu_1)$ and covariance $\Sigma_{22} - \Sigma_{21}\Sigma_{11}^{-1}\Sigma^{12}$ of $X_2|X_1$ satisfy linear regression, we can re-express

$$X_2|X_1 \sim \mathcal{N}(\mu_2 + \Sigma_{21}\Sigma_{11}^{-1}(X_1 - \mu_1), \Sigma_{22} - \Sigma_{21}\Sigma_{11}^{-1}\Sigma^{12}) \quad (6)$$

spreading the identity $J\Sigma = I$ to the following useful expression:

$$\begin{pmatrix} J_{11} & J_{12} \\ J_{21} & J_{22} \end{pmatrix} \begin{pmatrix} \Sigma_{11} & \Sigma_{12} \\ \Sigma_{21} & \Sigma_{22} \end{pmatrix} = \begin{pmatrix} I & 0 \\ 0 & I \end{pmatrix} \quad (7)$$

We re-express the Eq. (7) in terms of J:

$$X_2|X_1 \sim \mathcal{N}(\mu_2 - J_{22}^{-1}J_{21}(X_1 - \mu_1), J_{22}^{-1}) \tag{8}$$

Therefore, the missing entries of case i, $x_{mis,i}$ given $x_{obs,i}$ will also follow a normal distribution and the missing values satisfied the Eq. (9):

$$\hat{x}_{mis,i} = \hat{\mu}_{mis} - (\hat{J}_{mis,mis})^{-1}\hat{J}_{mis,obs}(x_{obs,i} - \hat{\mu}_{obs}) \tag{9}$$

We try to solve the problem in Eq. (5) through the Expectation Maximization(EM) algorithm. We assumed the data x have sufficient statistics number.

$$T_1 = x^T 1 = (\sum_{i=1}^{n} x_{i1}, \sum_{i=1}^{n} x_{i2}, \ldots, \sum_{i=1}^{n} x_{ip})$$

and

$$T_2 = x^T x = \begin{pmatrix} \sum_{i=1}^{n} x_{i1}^2 & \sum_{i=1}^{n} x_{i1}x_{i2} & \cdots & \sum_{i=1}^{n} x_{i1}x_{ip} \\ \sum_{i=1}^{n} x_{i2}x_{i1} & \sum_{i=1}^{n} x_{i2}^2 & \cdots & \sum_{i=1}^{n} x_{i2}x_{ip} \\ \vdots & \vdots & & \vdots \\ \sum_{i=1}^{n} x_{ip}x_{i1} & \sum_{i=1}^{n} x_{ip}x_{i2} & \cdots & \sum_{i=1}^{n} x_{ip}^2 \end{pmatrix}$$

The penalized negative log-likelihood Eq. (3) present in measure of the statistics T_1 and T_2 as follows:

$$-l(\mu, J; x) + \lambda\|J\|_1 = -\frac{n}{2}\log|J| + \frac{n}{2}\mu^T J\mu - \mu^T JT_1 + \frac{1}{2}tr(JT_2) + \lambda\|J\|_1 \tag{10}$$

From this T_1 and T_2 are linear relationship. Then the expected negative penalized log-likelihood was expressed as:

$$Q(\mu, J|\mu', J') = -\mathbb{E}[l(\mu, J; x)|x_{obs}, \mu', J'] + \lambda\|J\|_1 \tag{11}$$

Iterating the E step and M step, we get the missing entries corresponds the diagonal elements of covariance matrix. The iteration time m related the parameter $(\mu^{(m)}, J^{(m)})$.

E step: At this step, obtain the negative penalized log-likelihood expected value, T_1 and T_2 is linear relationship due to the complete penalized log-likelihood. The detail formulation of E step:

$$T_1^{(m+1)} = \mathbb{E}[T_1|x_{obs}, \mu^{(m)}, J^{(m)}] T_2^{(m+1)} = \mathbb{E}[T_2|x_{obs}, \mu^{(m)}, J^{(m)}]$$

The conditional expectation of x_{ij} and $x_{ij}x_{ij'}$ for $i = 1, \ldots, n$ and $j, j' = 1, \ldots, p$ is essential to estimate the above formulation. We assumed each case the set of missing entries and observed entries is multivariate normally distributed, Eq. (9) can be expressed as:

$$\mathbb{E}[x_{ij}|x_{obs}, i, \mu^m, J^m] = \begin{cases} x_{ij} & if \ x_{ij} \ is \ observed \\ c_j & if \ x_{ij} \ is \ missing \end{cases}$$

The vector c we defined as

$$c = \mu_{mis}^m - (J_{mis,mis}^m)^{-1} J_{mis,obs}^m (x_{obs,i} - \mu_{obs}^m)$$

Here, $J_{mis,mis}$ is the sub-matrix of J, each elements associated with the missing entries i. $J_{mis,obs}$ is analogously with $J_{mis,mis}$.

M step: At this step, maximized the expected log-likelihood through update mean vector and covariance matrix of multivariate normally distributed iteration by iteration. The Eq. (10) with the respect to μ approaching to 0, we obtain the update μ equation as:

$$\mu^{m+1} = \frac{1}{n} T_1^{m+1}$$

Re-organization the Eq. (10) through the following optimization equation to update the covariance matrix.

$$J^{m+1} = \underset{(\mu,J):J\succ0}{\arg\min} \left(-\log|J| + tr(JS^{m+1}) + \frac{2\lambda}{n}\|J\|_1\right)$$

where $S^{m+1} = \frac{1}{n} T_2^{m+1} - \mu^{m+1}(\mu^{m+1})^T$. From the above equation, the M step become to a standard GLasso problem. The expectation maximization algorithm step until convergence to stop iteration. The concentration matrix J diagonal elements associate with missing entries i. There we rank the diagonal elements in descending order according to their values, and select the top k values entries constructing a rank list.

3.2   Rank Aggregation of Selection Entries

This section we present the rank aggregation to ensure the important position of each case. The overarching idea is that we treat each case the top k uncertainties values with the covariance matrix of diagonal elements corresponding missing entries is a rank list l_i. what we want to do is from this list to ensure the important position of incomplete matrix for predicting.

The Shannon Entropy is use to measure the uniformness of the distribution of discordance. For a message, it also can be measure information content. Kemeny Optimal Aggregation is a good choice to reward a candidate entries of rank list which have a higher degree of uniformness in the associated distribution of disagreement. For mathematical to measure, Shannon Entropy give the uncertainty value corresponding with a random model. A random variable X, have finite amount of possible values $x_1, x_2, \ldots, x_n$. Suppose that p indicates the probability mass function, so Shannon Entropy detail formulation as follow:

$$Entropy(X) = -\sum_{i=1}^{n} p(x_i)log_{base}p(x_i) \tag{13}$$

Obviously, what we want is to find a consensus ordering l_i. where $|S| = k$, we have n input orderings $l_1, l_2, \ldots, l_n$ of k objects in S. The input list l_m is

$p(l_m)$ are too high probability to be incurred. Based on this, we express the formulation as:

$$p(l_m) = \frac{\tau(l_m, l_c)}{\sum_{i=1}^{n} \tau(l_i, l_c)} \tag{14}$$

l_c denotes a candidate optimal ordering. Therefore, we obtain the uniformness in the distribution of discordance of l_c through below formulation:

$$Entropy(l_c|l_1, l_2, \ldots, l_n) = -\sum_{j=1}^{n} p(l_j)log_{base}p(l_j) \tag{15}$$

So we scale $\tau(.,.)$ when separating it by $Entropy(.|.)$ and reformulate the Kemeny Optimal Aggregation as follows:

$$l^* = \underbrace{\arg min}_{l_c} \frac{\sum_{i=1}^{n} w_i \times d(l_c, l_i)}{Entropy(l_c|l_1, l_2, \ldots, l_n)} \tag{16}$$

when the $Entropy(.|.)$ increasing, above formulation give a candidate list similar to the optimal ranking.

3.3 Active Multivariable Matrix Completion

We are now ready to summarize the *Active Multivariate Matrix Completion*. As show in Algorithm 1, *Active Multivariate Matrix Completion* give us a straight way to complete the multivariate matrix. The major advantage of *Active Multivariate Matrix Completion* is that all queries from rank list entropy values, and each queries position is very important for prediction. Moreover, it is very persuasiveness. Because the number of queries is less than missing values, *Active Multivariate Matrix Completion* will consume less human oracles and aggregation budgets than traditional active matrix completion.

Active Multivariate Matrix Completion aims to actively ask a large amount of queries condition on missing values dataset with limited human oracles and aggregation budgets. In summary, this algorithm has the following features:

- First, we develop the single variate matrix problem to multivariate matrix problem. It's not only applies on the multivariate problems, but more importantly, it give a straight way to combine more influence factors for prediction.
- Second, Our mechanism decreases the noise of whole data samples. The active multivariate matrix completion combine rank aggregation to selected entries for human oracles.

Algorithm 1. Active Multivariate Matrix Completion

Require: Incompletely observed multivariate matrix M, set ϕ of observed entries, query size k, batch size s and amount of iterations n.

    for $iterations = 1 \rightarrow n$ **do**

        Transform the partially observed multivariate matrix into several single variate matrices

        For each matrix, compute the prediction of missing entries corresponding covariance matrix

        Sort the covariance matrix diagonal elements in descending order base on their uncertainty (variance) values

        For each case (row or column), the top k entries is a rank list l_i and do rank aggregation

        From human oracles to give the ground truth values of bottom s uncertain entries and save their position of bottom s uncertain entries

        Update the matrix through each iterations with newly acquired entries

        Using any standard matrix completion algorithm to complete the matrix

        Transform several single variate matrices into multivariate matrix

        Given the reconstruction error

    end for

    return The complete matrix

---

4   Experiment and Analysis

We discuss the performance of active multivariate matrix completion algorithms in this section. We manually deleted the entries data values at random percentage (ranging from 40% to 98%). The active multivariate matrix completion mainly focus on query a number of entries from rank aggregation given in each iteration. After the query process, human annotation with the selected locations in the incompletely matrix. (here we supply the ground truth value of the missing entries to simulate the human annotation). Matrix completion through any standard algorithm (in this paper our work base on SVD completion algorithm). Frobenius norm is adopted to compute the reconstruction error (the difference of the given data matrix and the predicted matrix). We iterate the procedure over and over. With the amount of iterations increasing or amount of observed entries increasing, the reconstruction error is decreasing. Compare the results against with the case merely using the SVD completion algorithm (referred in the below as passive completion) and where there was no human intervention, the case where the human annotations were selected at random (referred in the below as random), the case where just active matrix completion without rank aggregation, the case where use the mask graph to decide which position is important and then human annotations the selected entries (referred in the below as order extend).

4.1   Datasets

The experiments conducted on images dataset. We use four images with the size 256×256 —the Female, House, Tree and Jelly beans for our research. The images datasets are presented in Fig. 1. Percentage of missing entries set as 60%. Begin query size (rank list entries amount at each iteration) set as 80, the batch size (queried entries amount at each iteration) set as 40. This paper the procedure of iteration time is 40. The dataset was run 5 times through 5 algorithms (where we randomly select the missing entries at manual deleted entries matrix). Due to the datasets are relatively small, the standard SVD method is adopted. We use the proposed approach by [15] to estimate the covariance matrix. The results are presented in Fig. 2.

Female   House   Tree   Jelly beans

Fig. 1. RGB images used in our experiments

4.2   Experiments Analysis

Compare our algorithm performance with two state-of-the-art active matrix completion algorithms, *Order-Extend* and *AMC*.

Order-Extend: A linear systems of actively select entries for query the ground truth, introduced by [12]. The algorithm need to identify the singular value threshold, in our experiments we set singular value threshold is 5.

AMC: The first algorithm research active matrix completion, introduced by [11]. The algorithm based conditional normal distribution. Base on the results, the queried entries amount was expressed as x axis, the matrix reconstruction error was expressed as y axis. Obviously, with the more and more entries labeled the rate of the reconstruction error will decrease.

The top of dotted horizontal line denotes without human annotation to do matrix completion. SVD algorithm was selected to completed matrix completion. (emphasize that the observed entries amount remains the same). From the Fig. 2 we can see the reconstruction error of active multivariate matrix completion is obvious reduction than others. Furthermore, all the rest active matrix completion algorithm outperform than random selected entries annotation. With the observed entries amount increasing, the reconstruction errors drop speed faster. Further empirical researches point when the images remains same sparsity percentage sparsity (20%, 40% and 80%) - only the reconstruction errors different.

Then, we analyse the computation time of experiments matrix completion algorithms with human annotation. Table 1 display the results. The most efficient algorithm in computation time is Order-Extend, other than Random Sampling.

From the table the AMC (active matrix completion) and AMMC (active multivariate matrix completion) handle about 3.67 million entries spending approximately 5 min. Similarly, the Order-Extend handle same entries spending time less than 2 min. Comparatively, the AMMC algorithms error reduction are more efficient computationally than others. So that have the great potential to handle large datasets.

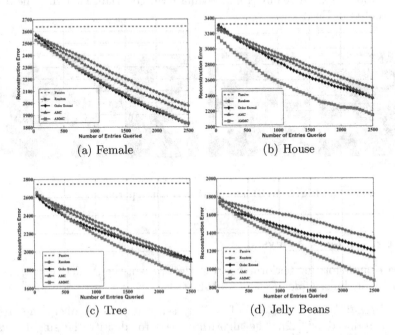

(a) Female   (b) House

(c) Tree   (d) Jelly Beans

Fig. 2. AMMC on image datasets. Degree of sparsity = 60%

Table 1. Average time taken to asking entries from the incompletely matrix.

Datasets	Random	Order-Extend	AMC	AMMC
Female	0.91	7.41	17.25	16.89
House	0.94	7.29	12.57	14.34
Tree	0.42	8.71	14.38	16.29
Jelly Beans	0.89	17.50	17.56	19.09

5   Conclusions

Active matrix completion is one influential notion in the matrix completion community. However, the proposed active matrix completion algorithms mainly focus on active selecting entries, failing to find the selected entries values with the location relationship. This paper presented a algorithm correlated rank aggregation have the following contributions:

- Correlated multivariate matrix with single variate matrix, give us a new sight for RGB image matrix completion, rather than use gray image recovering matrix.
- Rank aggregation is taken to find the relationship of the selected entries values with its location. Compared to the AMC and Order-Extend, the reconstruction error smaller.
- Selected entries location is a important information for predicting when the matrix have similar structure. Especially, for the very large scale datasets, it will save considerable computation time and human oracles.

References

1. Yick, J., Mukherjee, B., Ghosal, D.: Comput. Netw. **52**(12), 2292 (2008). https://doi.org/10.1016/j.comnet.2008.04.002
2. Candès, E.J., Tao, T.: IEEE Trans. Inf. Theory **56**(5), 2053 (2010). https://doi.org/10.1109/TIT.2010.2044061
3. Yuan, M., Lin, Y.: Biometrika **94**(1), 19 (2007). https://doi.org/10.1093/biomet/asm018
4. Candès, E.J., Recht, B.: Commun. ACM **55**, 111 (2012)
5. Kang, Z., Peng, C., Cheng, Q.: Proceedings of the Thirtieth AAAI Conference on Artificial Intelligence, 12–17 February 2016, Phoenix, Arizona, USA, pp. 179–185 (2016)
6. Zhang, X., Yin, C.: 9th International Conference on Wireless Communications and Signal Processing, WCSP 2017, 11–13 October 2017, Nanjing, China, pp. 1–5 (2017)
7. Krishnamurthy, A., Singh, A.: Advances in Neural Information Processing Systems 26: 27th Annual Conference on Neural Information Processing Systems 2013. Proceedings of a Meeting Held 5–8 December 2013, Lake Tahoe, Nevada, United States, pp. 836–844 (2013). http://papers.nips.cc/paper/4954-low-rank-matrix-and-tensor-completion-via-adaptive-sampling
8. Silva, J.G., Carin, L.: The 18th ACM SIGKDD International Conference on Knowledge Discovery and Data Mining, KDD 2012, 12–16 August 2012, Beijing, China, pp. 325–333 (2012). https://doi.org/10.1145/2339530.2339584
9. Wauthier, F.L., Jojic, N., Jordan, M.I.: The 18th ACM SIGKDD International Conference on Knowledge Discovery and Data Mining, KDD 2012, 12–16 August 2012, Beijing, China, pp. 1339–1347 (2012). https://doi.org/10.1145/2339530.2339737
10. Raymond, R., Kashima, H.: Machine Learning and Knowledge Discovery in Databases, European Conference, ECML PKDD 2010, 20–24 September 2010, Barcelona, Spain, Proceedings Part III, pp. 131–147 (2010). https://doi.org/10.1007/978-3-642-15939-8_9
11. Chakraborty, S., Zhou, J., Balasubramanian, V.N., Panchanathan, S., Davidson, I., Ye, J.: 2013 IEEE 13th International Conference on Data Mining, 7–10 December 2013, Dallas, TX, USA, pp. 81–90 (2013)

12. Ruchansky, N., Crovella, M., Terzi, E.: Proceedings of the 21th ACM SIGKDD International Conference on Knowledge Discovery and Data Mining, 10–13 August 2015, Sydney, NSW, Australia, pp. 1025–1034 (2015). https://doi.org/10.1145/2783258.2783259
13. Mavroforakis, C., Erdös, D., Crovella, M., Terzi, E.: Proceedings of the 2017 SIAM International Conference on Data Mining, 27–29 April 2017, Houston, Texas, USA, pp. 264–272 (2017). https://doi.org/10.1137/1.9781611974973.30
14. Städler, N., Bühlmann, P.: Stat. Comput. **22**, 219 (2012). https://doi.org/10.1007/s11222-010-9219-7
15. Mazumder, R., Hastie, T.: J. Mach. Learn. Res. **13**, 781 (2012). http://dl.acm.org/citation.cfm?id=2188412

Model of Decision Making for Agents Based on Relationship Coefficient

Zheng Wang[1(✉)] and Jiang Yu[2]

[1] College of Computer Science, Xi'an Shiyou University,
Xi'an 710065, Shaanxi, China
wangzheng@xsyu.edu.cn
[2] College of Printing and Packaging Engineering,
Xi'an University of Technology, Xi'an 710048, Shaanxi, China

Abstract. This paper is concerned with the problem of interactions among agents and how agents make decisions under this framework. Researchers contribute a lot to various interactions involved in agents and propose models of decision making process to portray how the agents behave. Based on relationship coefficient, a framework of interaction among agents is presented which is ranging from collaboration to antagonism. Furthermore, a model of decision making is put forward according to this framework. This model is illustrated and contrasted with an instance of game theory.

Keywords: Multi-agent system · Universal logic
General correlation coefficient

1 Introduction

The researches on autonomous agents and Multi-Agent System (MAS) have aroused intense interesting and are a comparatively new and multi-disciplinary subject [1]. The study of the basic theories and applications involves computer science, economics, philosophy, sociology and ecology. Because of the opening and dynamic, it provides a wider realm to probe into and a new perspective. There are various interactions within human society. Examples of common types of interactions include: cooperation, coordination and negotiation [1]. There are some other types of interaction which are not included in these examples, such as competition and antagonism. As we can consider the interactions among agents similar with those in human society, that means the study of interaction should include the range from complete cooperation to complete hostility. Many researchers pay much attention to the protocols, models, processes of interactions under different condition. They overlap in the research scope and isolate on the method.

Stone [2] divides the MAS interaction into two major categories, benevolent and competitive, and discussed it mainly in the perspective of co-evolution and learning. Zhang [3] measures off the MAS cooperation depending on the agents' goals and the degree of cooperation, ranging from complete cooperation to complete selfishness. Start from the sociality of agents, Jennings [4] proposes the socially responsible agents which are able to benefit from interactions with other agents, but are nevertheless

Q. Chen et al. (Eds.): ATIS 2018, CCIS 950, pp. 199–207, 2018.
https://doi.org/10.1007/978-981-13-2907-4_17

willing to provide some resources for the benefit of the overall system some of the time. The agents' actions can be classified by the comparison between value of personal benefit and that of social benefit. There are four kinds of actions defined: social actions, individual actions, divided actions and futile actions. Further on this work, Kalenka [5] observes a spectrum of potential decision making functions which ranges from the purely selfish (F_0) to the purely altruistic (F_n).

However, in either case, the classification of interaction is indistinct, and the decision making is rigid and isolate. There is not an explicit measure to include all sorts of interactions and tidy up the relationship among them. For example, how to describe a MAS in which some agents are cooperative while others are competitive or even antagonistic? Therefore, it is important and necessary to bring forward a clearer way to generalize all kinds of the interaction among agents. To achieve this, we introduce a relation coefficient and construct a framework of interaction by which we can explicitly depict the relationship between agents and tease out the common threads that make up the MAS interaction tapestry. Moreover, we propose a decision-making model for agents within this framework.

We begin in the following section by introducing the relation coefficient and constructing a framework of interaction. In Sect. 3, a decision-making model is proposed and a detailed explanation is given. Section 4 illustrates the use of the model on an example of game theory. In the end, comes the summary.

2 Interaction Framework for MAS Based on Relationship Coefficient (RC-h)

From the point of view of philosophy, nothing is completely isolated. There are various relationships between objects in the world. An entity would make a plan considering the relationship with others more or less. For instance, people belong to different organizations or play different roles in daily life and have different relationship with others. They could be relatives, friends, colleagues, adversaries, enemies or strangers. These could be abstracted as cooperation, selfishness, competition, antagonism and so on. These relationships can be considered as the basis of behavior decision. "Attitude is everything". In the same way, agents interact with each other according to the relationship between them.

Therefore, how to describe and tidy up these relationships is the primary issue of establishing a decision making model. He [6, 7] brings up the concepts of Generalized Correlativity and Generalized Correlation Coefficient. Generalized correlativity consists of consistent and exclusive relation while a coefficient called generalized correlation coefficient (h) is used to portray the ups and downs of the relations. Consistent relation includes affinity and repulsion while exclusive relation involves execution and viability.

We introduce these concepts into the interactions among agents and use h as the relationship coefficient (RC) to depict the various relationships between agents. For the purpose of simplification and convenience, the domain of the relationship coefficient is extended to $[-1, 1]$.

Briefly speaking, the interaction of agents can be classified into two major divisions: cooperation and non-cooperation. Correspondingly, the different range of value of h can be used to describe it, i.e. $h \in [0, 1]$ and $h \in [-1, 0]$ separately. Within the cooperation ($h \in [0, 1]$), it can be subdivide into collaboration ($h \in [0.5, 1]$) and selfishness ($h \in [0, 0.5]$) according to the desire to cooperate or be selfish. In collaboration, agents pay more attention to cooperate with other agents to fulfill goals together while in selfish situation, agents prefer to maximize their own profits. Similarly, in non-cooperation ($h \in [-1, 0]$), there are competition ($h \in [-0.5, 0]$) and antagonism ($h \in [-1, 0.5]$) according to the degree of contradiction and protection. In the competition, agents try their best to protect themselves, while in antagonism agents are more bellicose and aggressive. Figure 1 shows the relation between coefficient h and the interaction among agents.

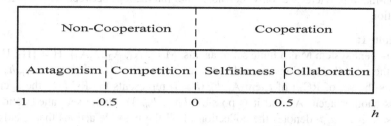

Fig. 1. This shows a fig imaging the relationship between RC-h and MAS interaction

The continuous change of h from 1 to −1 indicates the interaction changes of agents from collaboration to antagonism, passing through selfishness and competition. There are several particular points which attract more attention during this course. ① $h = 1$, complete collaboration. It is like the instances in Distributed Problem Solution (DPS). In such situation, agents have no local goal, they work together to solve the problem. They must cooperate with each other to achieve the global goal. Their cooperation is set in designing process. ② $h = 0$, complete selfishness. These agents are self-interested or rational. They perform to maximize their own profits. Many articles can be found in research on game theory which is a typical example. The main issues of rational agents are conflicts detecting and resolving. ③ $h = -1$, complete antagonism. In such condition, agents aim at destroying the adversary. Zero-sum is one of these paradigms. Besides, $h = 0.5$ and $h = -0.5$ are two other particular statuses. $h = 0.5$ is the boundary between selfishness and collaboration, while $h = -0.5$ is between competition and antagonism.

It can be seen this framework includes all kind of possible kinds of interactions among agents. Given such an interaction framework, we can delve the interaction systematically and tidy up the concepts and relationships within it clearly which were isolated and incomplete before.

3   Decision Making Model for Agent Based on RC-*h*

In an open and dynamic environment, the behaviors of situated agents show more complexity because of the autonomy, heterogeneity and diversity. The agents have so different structures, different statuses and different targets that it is rigid if they make decision according to same and inflexible strategy. When anonymous agents behave, they prefer to choose flexible action to adapt the changing of environment, objects and aims.

In our interaction framework mentioned above, all kinds of interactions are considered and included, from complete cooperation to complete antagonism. Now we present a decision-making model for agents based on this framework.

In order to discuss the interaction and decision-making model more explicitly, we make some necessary definition for MAS and agents, and then comes the model of decision-making. Here we only consider the interaction between two agents for facilitation.

Definition 1:
A multi-agent system M is a finite set of agents. $M = \{A_1, A_2, ..., A_n\}$. $H = \{H_1, H_2, ..., H_n\}$ is the set called relationship coefficient set, where $H_i = \{h_{i1}, h_{i2}, ...h_{i(i-1)}, h_{i(i+1)}, ...h_{in}\}$ is the set of RC-*h* of agent A_i, h_{ij} $(i \neq j)$ represents the RC-*h* of the agent A_i towards another agent A_j, and it is possible $h_{ij} \neq h_{ji}$. The finite set called action set $Act = <a_1, a_2, ...a_m>$ denotes the collection of all the possible actions that agents can perform. For simplification, we assume that all the agents have the same action set. a^* $\in Act$ represents the final action that agent choose to perform after decision making.

In our decision-making model, when agents choose which action to perform, they consider not only the effect of the action on their own, but also on the other participant agent. So we introduce a vector to describe both of the effects.

Definition 2:
Impact vector $I(a_k) = (I_1(a_k), I_2(a_k))$ represents the impacts of action $a_k \in Act$ $(k = 1, 2, ..., m)$ on the agent who performs it $(I_1(a_k))$ and the one who undertakes it $(I_2(a_k))$. Additionally, $I_1(a_k), I_2(a_k)$ are real number, and $1 \geqslant I_1(a_k), I_2(a_k) \geqslant -1$, where the positive value means the action does good to the agents while negative value means the action does harm to the agents, and 0 means no difference.

As we described above, there are various relationships between agents. Such relationships definitely influence the agents to make decision. That is when they are making decision they pay different attention to themselves and other participants depending on the relationships between them. Here, weighting factors are used to depict the different emphases on the impacts of action.

Definition 3:
Weighting vector $\Omega(h_{ij}) = (\omega_1(h_{ij}), \omega_2(h_{ij}))$ denotes the weighting factors added to $I(a_k)$, where $\omega_1(h_{ij}), \omega_2(h_{ij})$ imply the concern degree of the agent who perform the action on itself and on other's and defined as following. It is function of h_{ij}.

$\omega_1(h_{ij}), \omega_2(h_{ij})$ are depending on RC-*h* which denotes that the different relationship between agents determine the different behaviors. We have:

Definition 4:
$\omega_1(h_{ij})$, $\omega_2(h_{ij})$ which are in $\mathbf{\Omega}(h_{ij})$ are decided by

$$\begin{cases} \varpi_1(h_{ij}) = 1 - |h_{ij}| \\ \varpi_2(h_{ij}) = h_{ij} \end{cases} \tag{1}$$

When $h \leqslant 0$ the negative value of ω_2 is to distinguish the non-cooperation from cooperation. The relationship between h and ω_1, ω_2 is shown in Fig. 2.

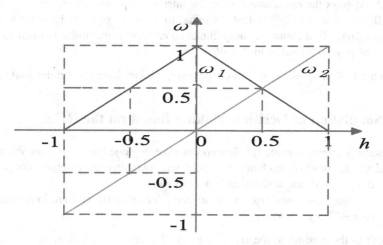

Fig. 2. This shows a figure depicting the relationship between RC-h and the weight $\omega 1$, $\omega 2$

So, when agents make decision, they take the whole impacts into account.

Definition 5:
The total impacts of the action a_k are described by:

$$\text{SI}(h_{ij}, a_k) = \mathbf{\Omega}(h_{ij})\, \mathbf{I}(a_k) = \omega_1(h_{ij})\, \mathbf{I}_1(a_k) + \omega_2(h_{ij})\, \mathbf{I}_2(a_k) \tag{2}$$

Definition 6:
The action $a^* \in \text{Act}$ which the agent finally chooses to perform after decision making is the one which maximize the total impact $\text{SI}(h_{ij}, a_k)$, that is

$$a^* = \arg\max_{a_k \in Act} SI(h_{ij}, a_k), k = 1, 2, \ldots, m \tag{3}$$

The analysis of this model is as follows considering the formulas (1), (2) and (3) with the change of h_{ij}, and concentrate on several particular values of h_{ij}. To simplify, let A_1 be the agent who makes decision and A_2 be the agent with which A_1 interacts. h represents h_{12}.

1. $0 \leqslant h \leqslant 1$, A_1 is cooperative. When h decreases from 1 to 0, A_1 changes its state from cooperation to selfishness. Meanwhile, $\omega_1(h)$ which is the weight to the influence on itself increases from 0 to 1 continuously while $\omega_2(h)$ which is that on

the other decreases from 1 to 0. These two processes mean A_1 pays more and more attention to the impact of actions on itself (i.e. $I_1(a_k)$) and pays less and less attention to that on A_2 (i.e. $I_2(a_k)$).

2. $h = 1$, A_1 is completely cooperative. At this point, $\omega_1(h) = 0$, $\omega_2(h) = 1$, no matter how much $I_1(a_k)$ is, $\omega_1(h)I_1(a_k) = 0$. It implies A_1 only thinks about how to maximize the benefit of A_2 and pays no attention to its own benefit. It's possible that A_1 selects an action which would damage itself.

3. $h = 0.5$, A_1 is cooperative and also self-interested. At this point, $\omega_1(h) = \omega_2(h) = 0.5$. A_1 pays the equal attention to the interest of A_2 and itself.

4. $h = 0$, A_1 is completely selfish. Similar to item 2, $\omega_1(h) = 1$, $\omega_2(h) = 0$ and $\omega_2(h) I_2(a_k) = 0$, it means A_1 only thinks about how to maximize the benefit of its own and pay no attention to that of A_2.

When $-1 \le h \le 0$, A_1 is non-cooperative, we can have the similar analysis.

4   A Simulation of Decision Making Based on the RC-h

The prisoner's dilemma game is a famous example in game theory. It is an abstraction of social situations where each agent is faced with two alternative actions cooperating and defecting. The detail description is as followed:

Two persons have been arrested as suspects for criminal, and are interrogated in separate rooms. They are told

1. If both of them admit to the crime, they get 2 years of imprisonment
2. If both do not admit to the crime, they can only get 1 years each because of lack of evidence
3. If one of them admits and the other does not, the defector is released while the other serves 3 years.

Now the payoff matrix of this dilemma is showed in Fig. 3, where "cooperate" means not to admit to the crime, and "defect" means to admit. The values in the payoff matrix are not the numbers of the years of imprisonment, but express how good the result would be for each agent. The number lies in left-up corner of each cell is the payoff of Agent-A, and the one in right-down corner is that of Agent-B.

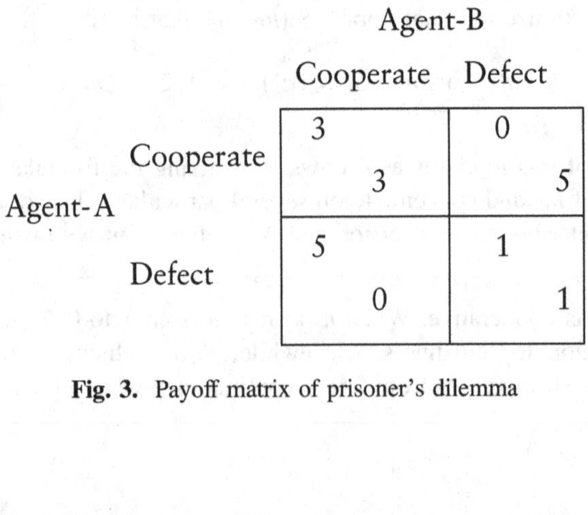

Agent-B

	Cooperate	Defect
Cooperate	3 3	0 5
Defect	5 0	1 1

Agent-A

Fig. 3. Payoff matrix of prisoner's dilemma

From Fig. 3, we can tell there are four strategies: s_1(C,C), s_2(C,D), s_3(D,C) and s_4(D,D), where C means Cooperation and D means Defection. The first parameter of each strategy represents the action of agent-A, the second one denotes that of agent-B.

In the analysis of game theory, s_4(D,D) is called the dominant strategy and also the Nash equilibrium. However, it is obviously not the best choice. For if each agent choose to cooperate, the final strategy would be s_1(C,C), both utility is 3 which is higher then 1 when they decide to defect. The reason of such dilemma is the conflict between personal rational and social rational. Agents always prefer to the strategy which can bring them the maximum profit so as to controdict with that of collectivity or society. In multi-agent system, the interactions are various in different scenarios. The rational asked in game theory has its limit. For instance, in the DSP, the agents are required to be altruistic and have no local target for themself but only work for the global target. Therefore, a more flexible and wider applicable decision making method is needed to solve such situation.

According to the decision making model based on RC-h mentioned above, an experimental simulation is designed. Figure 4 shows for agent-A and agent-B respectively, the total impacts (SI) of four strategies s_1(C,C), s_2(C,D), s_3(D,C) and s_4(D,D) calculated by formula (1). Because of the symmetry of payoff matrix, the curves of Agent-A are similar with those of Agent-B. Figure 5 shows the final chosen actions of Agent-A and Agent-B according to the decision making rule (formula (2), also refer to Fig. 4). This graph shows that when $h > 0.4$, both agents choose to cooperate. This value is near $h = 0.5$ which is the dividing lines of selfishness and collaboration. It is necessary to point out this turning point will drift with the difference of payoff matrix.

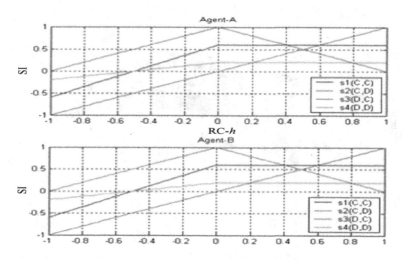

Fig. 4. The total impacts (SI) of the strategies of Agent-A and Agent-B

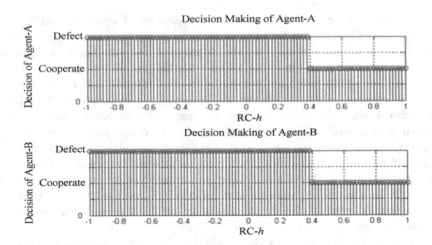

Fig. 5. The final decision of Agent-A and Agent-B according to RC-h

Figure 6 combines the results of Agent-A and Agent-B, and comes out the final decision which is effected by both sides. From Fig. 6 we can tell $h = 0.4$ is a boundary, when both agents' RC-h are less than it, the final result comes to $s_4(D,D)$ which coincide with the action of "rational" agents. What is different is the cooperative behavior ($s_1(C,C)$) occurs when the RC-h of two parties are greater than 0.4, i.e. both sides are nearly in the condition of collaboration. Else, $s_2(C,D)$ or $s_3(D,C)$ turns up if one agent is towards cooperative while the other is opposite.

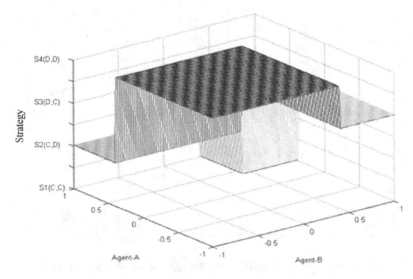

Fig. 6. The synthetic result of Agent-A and Agent-B

This decision making model also can be applied on other instances of game theory or other fields. We would not extend here.

From the analysis and simulation mentioned above, we can clearly tell the decision making model based on relationship coefficient-h is more flexible and reasonable. It allows us to comprehend the interactions in agents and its result from a higher view. Meanwhile, adjusting the RC-h according to design targets or learning the RC-h from interaction history makes the agents adaptive to the changing environment.

5 Summary

This paper presented a framework of interaction in MAS based on relationship coefficient-h. It provided a typology of the types of MAS interactions, ranging from complete cooperation to complete antagonism, and quantify the interaction by h. Within this framework, a decision-making model for agents is proposed which enable agents choose their action depending on the relationship to others. Using this model, in open and dynamic surroundings, agents can perform more adaptive and satisfy the design goals. At the end, the model is illustrated and contrasted with an instance of game theory.

References

1. Jennings, N., Sycara, K., Wooldridge, M.: A roadmap of agent research and development. Auton. Agents Multi-agent Syst. **1**(1), 7–38 (1998)
2. Stone, P., Veloso, M.: Multi-agent systems: a survey from a machine learning perspective. Technical report CMU-CS-97-193, Computer Science Department, pp. 181–217. Carnegie Mello University, Pittsburgh, PA (1997)
3. Zhang, W., et al.: Information Technology of Intelligent Cooperation. Publishing House of Electronic Industry, Beijing (2012)
4. Jennings, N.R., Campos, J.R.: Towards a social level characterization of socially responsible agents. IEE Proc. Softw. Eng. **144**(1), 11–25 (1997)
5. Kalenka, S., Jennings, N.R.: Socially responsible decision making by autonomous agents. In: Korta, K., Sosa, E., Arrazola, X. (eds.) Cognition, Agency and Rationality, vol. 79, pp. 135–149. Springer, Heidelberg (1999). https://doi.org/10.1007/978-94-017-1070-1_9
6. He, H., et al.: Principles of Universal Logics. Science Publishing Press, Beijing (2011)
7. He, H., Ai, L., Wang, H.: Uncertain and the flexible logics. In: ICMLC 2003 (2003)

Emotion Recognition in a Conversational Context

Binayaka Chakraborty[(✉)] and M. Geetha

Department of Computer Science and Engineering,
Manipal Academy of Higher Education, Manipal, India
binayakachakraborty@gmail.com,
geetha.maiya@manipal.edu

Abstract. The recent trends in Artificial Intelligence (AI) are all pointing towards the singularity, i.e., the day the true AI is born, which can pass the Turing Test. However, to achieve singularity, AI needs to understand what makes a human. Emotions define the human consciousness. To properly understand what it means to be human, AI needs to understand emotions. A daunting task, given that emotions may be very different, for different people. All these get even more complex when we see that culture plays a great role in expressions present in a language. This paper is an attempt to classify text into compound emotional categories. The proposal of this paper is identification of compound emotions in a sentence. It takes three different models, using Deep Learning networks, and the more traditional Naïve Bayes model, while keeping the mid-field level using RAKEL. Using supervised analysis, it attempts to give an emotional vector for the given set of sentences. The results are compared, showing the effectiveness of Deep Learning networks over traditional machine learning models in complex cases.

Keywords: Machine learning · Deep learning · RAKEL · Naïve Bayes
Sentiment analysis · Multi-labelled emotions

1 Introduction

Artificial Intelligence is the flavour of the current era that we are living in. From being touted as the Fourth Revolution, to emerging from the realms of Science Fiction into the real world. The challenge is to design and implement the computational mind, which was described by Descartes as a "universal instrument that is alike in every occasion" [1].

Drawing inspiration from nature, and moving towards a quantum computational future where events need not be limited to a binary expression, but instead may be described as a collection of expression, it is proposed that the current method of understanding sentiment be revised.

Affective Computing has been of attracting researchers in the current decade, as scientists seek to close the communication gap that is currently existing between computers and humans, being inspired from the unavoidable links that arise from emotions and human intelligence, having social interaction, making rational decision and many more.

© Springer Nature Singapore Pte Ltd. 2018
Q. Chen et al. (Eds.): ATIS 2018, CCIS 950, pp. 208–214, 2018.
https://doi.org/10.1007/978-981-13-2907-4_18

Creating intelligent agents with emotions has long been of interest to researchers. Many alternative models from psychological literature have been integrated into agent architectures, as in goal selection [2], belief generation [3], attention and learning.

Recently, an enormous effort is being made in the field of text based emotion detection, in part due to the constantly increasing amount of text available like chat logs and the truly enormous amount of social media content. The techniques most used in this type of recognition generally fall under three broad categories: (I) Machine Learning methods [4–6], (II) Emotional keyword spotting [7–9], and (III) Concept base methods [10, 11].

The methods of machine learning in this field are generally based on traditional text classification, given by an input text denoted by a bag-of-words model, and then various feature-extraction techniques, such as specific word counts, and identification of negation areas are applied that maps the text to a feature-vector matrix. Following this step, support vector machines [5] and conditional random fields [12] are applied to determine the type of the emotion the input text belongs to. However, such methods have achieved only modest performance in this domain [4].

The main reason appears to be a failure in capture of additional emotional information, i.e., beyond what is provided by the input text, by the traditional machine learning techniques. In an attempt to increasing the efficiency, weighted hidden order markov chaining technique was demonstrated [13] to obtain a better result.

Emotional keyword spotting methods depend on the availability of obvious emotional words. They are straightforward and easy to implement. The drawback of these methods is quickly visible whenever negation is involved, and they produce poor results.

Web ontologies or semantic networks are used in concept based methods, to analyse sentiment in text [14]. They are believed to be better than techniques which are purely syntactic because they can analyse multi-word expressions not expressly conveying emotions, but belong to concepts that show emotion.

Performance of concept based methods are directly proportional to the diversity and extensiveness of the semantic knowledge base, which vary enormously for different languages. The lexical resource SentiWordNet [15] was developed to support affective analysis of the natural language English.

SentiWordNet is the result of the automatic annotation of all synsets of WordNet [16], based on the notions of "positivity", "negativity" and "objectivity". Three numerical scores are associated with each synsets, indicating the aforementioned notions. Berardi et al., using SentiWordNet, recommended assigning a sentiment score for detecting the polarity (or sentimental value) of a given hyperlink [17]. Similarly, Baccianella et al., using SentiWordNet, established a system called StarTrack which could automatically rate product reviews based on how positive they were. It was done by constructing a word-level dictionary built upon SentiWordNet [15].

Taking inspiration from SentiWordNet, SenticNet [18] has been constructed for the exploration of polar information from colloquially used terms in everyday language to exhibit positive and negative viewpoints, such as "I have a bad feeling about this", or "She is on cloud nine these days".

In [19], the enormous taxonomy common knowledge base was combined with a natural-language based semantic network of common-sense knowledge, and the final knowledge base was the result of the application of multi-dimensional scaling, so than an open-domain sentiment analysis could be performed on it.

2 Classifiers Used

   i. Naïve Bayes
  ii. RAKEL
 iii. Deep Neural Network.

Naive Bayes is chosen as the baseline by which the performance of the other models will be judged. It is a classical model, and can give surprisingly good results provided the dataset contains enough independent variables for it to differentiate properly.

RAKEL (Random K-labelsets) is used because we have multiple label types for singular sentences. This, combined with Naïve Bayes, gives us a good idea about how traditional machine learning classifiers will perform in these kind of complex cases.

For the Deep Neural Network, we vary the configuration in order to test the effectives of the various architectures. The main issue with DNN is that the size of the dataset is very small, which is not ideal for creating DNNs. Varying the amount if hidden layers as well as the epochs should let us fine tune for a good performing model.

The loss function is defined as a linear regression head with mean squared loss (simple L2 loss).

$$MSE(\bar{X}) = E((\bar{X} - \mu)^2) \tag{1}$$

3 Dataset

The dataset used is provided by [20]. Out of the basic emotions listed in [21], the dataset provides the following labelled possible combinations of emotions (Table 1):

Table 1. Emotions in the dataset

Anger_Sadness	Fear_Interest_Love
Surprise	Joy_Sadness
Fear_Surprise	Joy
Love	Anger_Fear_Love
Interest_Joy	Interest_Love_Surprise
Joy_Surprise	Sadness
Interest	Fear
Joy_Love	Fear_Love_Surprise
Sadness_Surprise	Anger_Interest
Interest_Love	Interest_Surprise
Anger	Fear_Love
Fear_Love_Sadness	Interest_Sadness
Anger_Interest_Sadness	Love_Surprise

Examples:

(a) This film can be enjoyed on so many levels and I really enjoyed the third act. **Interest_Joy**

(b) It is art captured on film. **Love.**

Apart from these, if any sentence does not have any true identifiable emotion, it has been labelled as 'None'.

The entire spectrum of describable emotions has been taken from the following Fig. 1:

The words and their synonyms are used to create the training sets for the classifiers.

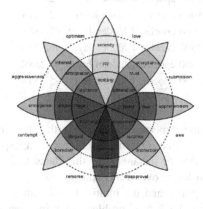

Fig. 1. R. Plutchik's entire range of emotions [Source: https://en.wikipedia.org/wiki/Contrasting_and_categorization_of_emotions#Plutchik's_wheel_of_emotions]

4   Experimental Analysis

The total number of samples in the training set is 503, while the test set contains 126 samples. The tools used for pre-processing the datasets are Eclipse and GSON package, in order to conform to a type of input TensorFlow can process. Taken as a whole, the deep learning network outperforms Naïve Bayes by a huge margin, but that margin depends on the number of epochs as well as the number of hidden layers chosen. RAKEL performs well as a mid-level technique, however, the overall average of RAKEL loses out to the DNN. This is likely because of the low density of training data, which doesn't let the classifier learn properly. Also, some label subsets, like 'Interest' occur only once in the training set. This can skew the learning method, making proper division improbable (Tables 2 and 3).

Table 2. Total dataset accuracy

Model	Accuracy
Deep learning (best result)	96.825%
RAKEL (averaged)	85.4%
Naïve Bayes	65.208%

Table 3. Deep learning individual accuracy

Model	Accuracy
1000 epochs, 5 hidden layer	68.25%
1000 epochs, 4 hidden layer	82.539%
500 epochs, 2 hidden layer	92.063%
1000 epochs, 2 hidden layer	96.032%
1500 epochs, 1 hidden layer	96.825%
2000 epochs, 1 hidden layer	95.238%
2500 epochs, 1 hidden layer	95.138%

It is observed from the experimental analysis that it is essential to design the network with a degree of caution. Overfitting is a problem, especially with the dataset being limited in the number of instances. In this dataset, the highest accuracy was determined with 1 hidden layer, over 1500 epochs.

In [20], the authors approached the multi-label classification problem in two different ways, (1) reducing the k-way problem to k-independent problem, so as to implement the solution as a SVM model [known as one vs rest technique], and (2) the random k-label sets (RAKEL) algorithm. The table below lists the difference between RAKEL and DNN the best deep learning method is given below (Table 4):

Table 4. Comparison between accuracies

DNN	RAKEL
.968	.854

5 Conclusion

It is observed that Deep Learning appears to be a better classification technique than traditional machine learning classification techniques, especially with such a small dataset size. RAKEL outperforms Naïve Bayes, but due to the small dataset and skewed distribution of label sets, doesn't manage to achieve the accuracy of DNNs. Also, the results in DNNs vary by a huge amount depending on the architecture used. The worst performance of the DNN is comparable to the accuracy of the Naïve Bayes classifier, due to the small dataset and the increased amount of hidden layers. It is seen that the low available data for each labelled doesn't lend well to an increase in hidden layers, as more layers tend to over-fit and bring down the overall accuracy.

Increasing the epochs also follows a kind of bell curve, as can be seen from the 500, 100 and 1500 values, with the epochs after 2000 tending to hover around 95%.

It is difficult to say with certainty the architecture needed to perform the classification for these type of experiment, as it varies wildly with the amount of annotated data available for training. The future prospect of this paper would be to test out the DNN architectures on a larger dataset.

References

1. Descartes, R.: Meditations on First Philosophy. Cambridge University Press, Cambridge (1631)
2. Fridja, N., Swagerman, J.: Can computers feel? Theory and design of an emotional model. Cogn. Emot. **1**(3), 235–257 (1987)
3. Marsella, S., Gratch, J.: A step towards irrationality; using emotion to change belief. In: Proceedings of the First International Joint Conference on Autonomous Agents and Multiagent Systems, Bologna, Italy (2002)
4. Mishne, G.: Experiments with mood classification in blog posts. In: First Workshop on Stylistic Analysis of Text for Information Access, Salvadore, Brazil (2005)
5. Teng, Z., Ren, F., Kuroiwa, S.: Recognition of emotions with SVMs. In: Huang, D.S., Li, K., Irwin, G.W. (eds.) ICIC 2006. LNCS, vol. 4114, pp. 701–710. Springer, Heidelberg (2006). https://doi.org/10.1007/11816171_87
6. Xia, R., Zong, C., Li, S.: Ensemble of feature sets and classification algorithms for sentiment analysis. Inf. Sci. **181**, 1138–1152 (2011)
7. Chungling, M., Predinger, H., Ishizuka, M.: Emotion estimation and reasoning based on affective textual interaction. Affect. Comput. Intell. **3784**, 622–628 (2005)
8. Hancock, J., Landrigan, C., Silver, C.: Expressing emotion in text-based communication. In: SIGCHI Conference on Human Factors, San Jose, California, USA (2007)
9. Li, H., Pang, N., Guo, S.: Research on textual emotion recognition incorporating personality factor. In: Proceedings of the International conference on Robotics, Crowne Plaza Sanya, Sanya, China (2007)
10. Grassi, M., Cambria, E., Hussain, A., Piazza, F.: Sentic web: a new paradigm for managing social media affective information. Cogn. Comput. **3**, 180–489 (2011)
11. Olsher, D.: Full spectrum opinion mining: integrating domain, syntactic and lexical knowledge. In: Sentiment Elicitation from Natural Text for Information, pp. 693–700 (2012)
12. Yang, C., Lin, K., Chen, H.: Emotion classification using web blog corpora. In: Proceedings of IEEE/WIC/ACM International Conference on Web Intelligence, Marriott Fremont, Fremont, CA, USA (2007)
13. Quan, C., Ren, F.: Weighted high-order hidden Markov models for compound emotions recognition in text. Inf. Sci. **329**, 581–596 (2016)
14. Cambria, E., Schuller, B., Xia, Y., Havasi, C.: New avenues in opinion mining and sentiment analysis. IEEE Intell. Syst. **28**, 15–21 (2013)
15. Baccianella, S., Esuli, A., Sebastiani, F.: SentiWordNet 3.0: an enhanced lexical resource for sentiment analysis and opinion mining. In: Proceedings of the Seventh Conference on Language Resources and Evaluation, LREC, Malta (2010)
16. Fellbaum, C.: WordNet: An Electronic Lexical Database. MIT Press, Cambridge (1998)
17. Berardi, G., Esuli, A., Sebastiani, F., Silvestr, F.: Endorsements and rebuttals in blog distillation. Inf. Sci. **249**, 38–47 (2013)

18. Cambria, E., Olsher, D., Rajagopal, D.: SenticNet 3: a common and common-sense knowledge base for cognition-driven sentiment analysis. In: Proceeding of AAAI, 2014, Palo Alto, California, USA (2014)
19. Cambria, E., Song, Y., Wang, H., Howard, N.: Semantic multi-dimensional scaling for open-domain sentiment analysis. IEEE Intell. Syst. **29**, 44–51 (2014)
20. Buitinck, L., van Amerongen, J., Tan, E., de Rijke, M.: Multi-emotion detection in user-generated reviews. In: Hanbury, A., Kazai, G., Rauber, A., Fuhr, N. (eds.) ECIR 2015. LNCS, vol. 9022, pp. 43–48. Springer, Cham (2015). https://doi.org/10.1007/978-3-319-16354-3_5
21. Plutchik, R.: The circumplex as a general model of the structure of emotions and personality. Am. Psychol. Assoc. **10**(1037), 17–45 (1997)

Machine Learning Based Electronic Triage for Emergency Department

Diana Olivia$^{(\boxtimes)}$, Ashalatha Nayak , and Mamatha Balachandra

Manipal Academy of Higher Education, Manipal, India
{diana.olivia,asha.nayak,mamtha.bc}@manipal.edu

Abstract. Increase in the number of casualties visit to Emergency Department (ED) have lead to over crowd and delay in medical care. Hence, electronic triaging has been deployed to alleviate these problems and improve managing the patient. In this paper research methodology framework based on diagnostic and cross-sectional study is used for patient triage. The empirical approach is used to build models for patient triage to correctly predict the patient's medical condition, given their signs and symptoms. Models are built with supervised learning algorithms. The "Naive Bayes", "Support Vector Machine", "Decision Tree" and, "Neural Network" classification models are implemented and evaluated using chi-square statistical test. This study infers the significance of using machine learning algorithms to predict patient's medical condition. Support Vector Machine and Decision Tree have shown better performance for the considered dataset.

Keywords: Medical triage · Emergency department
Supervised machine learning algorithms · Research methodology
Statistical analysis

1 Introduction

Increase in the number of casualties visit to Emergency Department (ED) have lead to the crowding and delay in the medical care. Subsequently leading to increased mortality, morbidity and poor process measures across the clinical conditions. Delay in care makes the critically ill patients more vulnerable to a worse health condition. At Emergency Department, triage procedure is used to identify the medical condition of the ill patients and decide further ED care based on the severity of their condition. Accurate triage process must be maintained at EDs to quickly identify and prioritize patients based on medical condition so that critical patients are given more priority compared to patients with less urgency. Even though the triage is a simple process but it is challenging due to the limited time, limited information varied medical conditions and completely rely on once perception.

The medical group generally applies a "Simple Triage And Rapid Treatment (START)" [1] method to access the patient's medical condition. The START

© Springer Nature Singapore Pte Ltd. 2018
Q. Chen et al. (Eds.): ATIS 2018, CCIS 950, pp. 215–221, 2018.
https://doi.org/10.1007/978-981-13-2907-4_19

protocol uses three physiological signs i.e. "pulse rate", "breathing rate", and "mental status" to sort the patients. Based on the conditions of these signs, patients are categorized into four groups which indicate the priorities of patients treated. The different groups and their corresponding treatment level are given as follows: "Green:" A patient with negligible injuries who do not require an immediate medical treatment. "Yellow:" A patient with significant injuries but treatment can be postponed for short period of time. "Red:" A patient with major injuries and need an immediate medical treatment, "Balck:" A patient who is deceased. Initially, all the casualties who can walk and follow the simple command are marked as a green category by the medical team. Later, based on the assessment of respiration and capillary refill pulse the casualty is marked as RED. The patient who is not breathing is marked as black. Over a time the casualty needs to be triaged again since the casualties medical condition may change over a time. Hence, the paper-based triage is not efficient which may not provide the latest medical condition of the casualty.

Recently, this problem is addressed by an automated electronic triage system, where vital sign sensing sensors are deployed on the human bodies to continuously monitor the vital sign parameters. Further, sensed information is analyzed by the medical team members using decision support system, to know the current medical status of the casualty. In this study, we applied machine learning algorithms to identify the medical status of the casualty. Our aim is to build a different machine learning models for identifying the casualties medical status and evaluate their performance and feasibility.

Based on above discussion the problem statement is defined as, to improve the quality of health care services during emergency situations, an efficient triage model is designed which will predict patient's medical condition. In this regard the Research Questions identified are, what machine learning models are applicable for assessment and prediction? and how to measure the effectiveness of the Models in prediction? Based on mentioned problem definition and research questions, the identified objectives are, Building Triage Models using Machine Learning algorithms for Patient Triage; Analyze and Compare built Models based on model's prediction accuracy; and Test the Effectiveness of the models on the prediction. The Rationale of study is to improve the quality of health service at the emergency department. Further, using the model it is possible to prioritize the patient based on their latest health condition. Hence the study is focused on assessing a patient medical condition by considering four algorithms namely "Decision Tree", "Support Vector Machine", "Neural Network", and "Naive Bayes". The paper is organized as follows. Section 2 discuss about the literature review. Section 3 about the methodology adopted, and Sect. 4 analyses different learning algorithms.

2   Literature Review

Caicedo-Torres et al. [6] have proposed the use of Learning algorithms in a "Pediatric Emergency Department (ED)" to predict which casualties should be admitted to the "Fast Track" based on their symptoms. The lean Thinking approach has been used to successfully handle the casualties. In this approach, EDs include separate patient stream called Fast Track, where low complexity people are treated, thus reducing the total waiting time of treating complicated casualties who are put under Regular Track. They have considered data set of 1205 instances with 74 variables. Admission to hospital is the response variable, with 70% of training examples fitting to the non-control group, i.e. discharged without hospitalization. Models were built to classify casualties to the Regular Track (positive class) and Fast Track (negative class). The performance of the classifier on the training set is validated using 10-Fold Cross-validation. Standard measures of Precision and Sensitivity and the F1-score are used to evaluate the model performance. The result shows that Multilayer Perceptron performs better compared to Support Vector Machine and Logistic Regression techniques. Prediction is only applicable to Pediatric patients.

Ghive et al. [7] have identified the diseases and corresponding treatments in text representation using NB and SVM classification models. Model performance is tested using Accuracy, Precision, F-measure, and Recall criteria. Experimental outcome specify that SVM performs better than NB. They have used Numerical and text form dataset for training and testing purpose. Various different classification techniques have been applied to the three different health care datasets (lung cancer Dataset, diabetes dataset, and hepatitis Dataset) taken from UCI repository. Dataset 2 is collected from Medline2 2001 abstracts. Sentences from titles and abstracts are annotated with entities and with eight relations. The implementation has been tested on MATLAB. The drawback of the work is considered for one specific disease.

Lee et al. [8] have classified breathing data into regular and irregular classes based on the breathing data features e.g., the amplitude of breathing cycle, vector-oriented feature, and breathing frequency. They have considered Neural network for the classification. The sensitivity and specificity of the proposed non-regular breathing pattern were analyzed. Receiver Operating Characteristic (ROC) curve analysis is used to validate the classification accuracy, and to obtain accurate value for the degree of irregularity. The proposed model has high sensitivity and specificity. True Positive Rate of proposed work is less than the existing work.

Claudio et al. [3] have proposed a "dynamic multi-attribute utility theory" based decision support system for prioritizing the patients in the emergency section. Data from 47 patients and 12 nurses are used for the study. The proposed MAUT model was evaluated by a physician who performed patient severity ranking decisions based on medical knowledge. A statistical analysis result shows not much difference between the suggestion proposed by the model and the medical decisions taken by the physician. The analysis result gives evidence for the benefit of merging technology with medical decision theory to support medical practitioner in prioritizing ED patients. The drawback of the work is patient prioritization list is not dynamic.

Elalouf et al. [4] have proposed Dynamic programming approach to schedule the patients at the emergency department. They have developed an algorithm by considering a basic set of assumptions and simulated it using data from the ED. The simulation study indicates that the "Floating Patients" method can reduce ED crowding and patient length-of-stay, without decreasing the quality of care. In Floating patients approach patients are sent to another hospital for examinations and treatment, to reduce their waiting time at the emergency department.

Salman et al. [5], have proposed a methodology for Triage and Prioritizing Chronic Heart Diseases patients through Telemedicine Environmental. Weights are assigned for individual vital signs and Severity index is computed using "TOPSIS (Technique for Order of Preference by Similarity to Ideal Solution)" method. The work is intended for Heart disease patients.

3   Methodology

In this section, research methodology adopted for the work is discussed. The purpose of the research is quantitative since quantitative data is considered for analysis. Data is searched in a systematic manner, hence method used for searching facts is scientific. Type of research is diagnostic and Experimental. The research approach is inferential. The empirical research method used is Prediction and Classification studies. The techniques used are Decision Support, Support Vector Machine and Neural Network. The study is cross-sectional. With experimental strategy, choices are multiple methods. The research approach is inductive in nature since from specific observations, broader decisions are made and model is developed. Research philosophy implied are positivism and functionalism. Positivism, since it believes that information obtained from logical and mathematical treatments, reports of sensory experience is the source of all authoritative knowledge. Functionalism is the principle that any model is built for the purpose. The method used for Data Collection is secondary Source. The data has eight numerical values which are considered as independent variables. The dependent variable is patient's condition which is normal or abnormal. The dataset has following attributes such as body temperature, Blood Pressure (Diastolic and Systolic), Blood Oxygen level, Respiration Rate, and Pulse Rate.

4   Result and Discussions

In this section, model comparison and data analysis are discussed. Models are built using Decision Tree, Support Vector Machine, Neural Network, and Naive Bayes learning algorithms. Algorithms are implemented using Python and Kerasa package. Perceptron Neural network model is constructed with 0.3 as learning rate and Sigmoid as activation Function.

To evaluate the model K-fold cross-validation technique is used, with K set to 10. In this technique, considered dataset is separated into K subsets. Further, training is repeated K-times. During each training, $K - 1$ subparts are considered for training and remaining one subset is used for testing. During each training iteration, a different one of these subsets is used as a validation set. The final accuracy of the model is computed using average across all K trails. As K increases the variance of the estimate is reduced. The classifier result may be positive or negative representing abnormal or normal patient' condition respectively. To measure the performance of the built models accuracy, sensitivity, specificity and precision criteria are used which is shown in Table 1. Sensitivity is defined as the probability that, the classifier result shows abnormal when in fact the patient condition is abnormal. Specificity is defined as the probability that the classifier result shows normal when in fact the patient condition is normal. Accuracy defines the probability of correct prediction and precision mention that portion of patients that are diagnosed as abnormal is actually abnormal. The classifier should have high specificity and sensitivity. The result shows that SVM and DT performs better.

Table 1. Performance metrics of the models.

Model	Accuracy (%)	Sensitivity (%)	Specificity (%)	Precision (%)
NN	60	87	18	61
NB	82	84	80	86
DT	84	87	81	87
SVM	84	87	81	87

The third objective, to check the effectiveness of the model in patient triage is done using statistical analysis. Hence, the considered alternate Hypothesis is 'Model has significant influence in correct prediction' and null Hypothesis is 'Model has no significant influence in correct prediction'.

Chi-square statistical test [2] is used to check the model effectiveness in patient medical condition prediction. The significance level is set to 0.05. The observed frequency values are computed. The expected chi-square values are shown in a Table 2, which is computed using observed values. Finally, chi-square value is computed using expected values by applying chi-square formula, which is 23.567.

Using standard chi-square table, the tabulated chi-square value for the mentioned DOF and significance level is 7.815. Finally, to test the acceptance of the above mentioned hypotheses tabulated and computed chi-square values are compared. The obtained chi-square value is more than tabulated chi-square value which shows rejection of the null hypothesis. Hence, statistical analysis showed that there is a significant effect of the model in patient triage process.

Table 2. Expected chi-square values

O_{ij}	E_{ij}	$O_{ij} - E_{ij}$	$(O_{ij} - E_{ij})^2/E_{ij}$
84	77.5	6.5	0.545
16	22.5	−6.5	1.877
84	77.5	6.5	0.545
16	22.5	−6.5	1.877
82	77.5	4.5	0.261
18	22.5	−4.5	0.9
60	77.5	−17.5	3.951
40	22.5	17.5	13.611

5   Conclusion

Paper presents patient triage using different machine learning techniques. Dataset containing patient vital sign information collected at the emergency department is considered for the experiment. Models are compared using sensitivity, specificity, precision and accuracy metrics. The comparison result shows that Support Vector Machine and Decision tree performs better in patient triage process for the considered dataset. Further, chi-square statistical analysis is performed to verify the effectiveness of the models in patient triage process. The analysis result shows that model has a significant effect on patient triage. In future, the work may be extended to triage the patient into more classes like critical and moderate. Further, additional vital sign and patient's statical information like ECG, age, gender could be considered to improve the accuracy of the patient triage process.

References

1. Sakanushi, K.H., et al.: Electronic triage system for continuously monitoring casualties at disaster scenes. J. Ambient Intell. Humani. Comput. **4**(5), 547–558 (2012)
2. Kothari, C.R.: Research Methodology, Methods and Techniques, 2nd edn. New Age International Limited, New Delhi (1990)
3. Claudio, D., Kremer, G.E.O.: A dynamic multi-attribute utility theory based decision support system for patient prioritization in the emergency department. IISE Trans. Healthc. Syst. Eng. **4**(1), 1–15 (2014)
4. Elalouf, A., Wachtel, G.: An alternative scheduling approach for improving emergency department performance. Int. J. Prod. Econ. **178**, 65–71 (2016)
5. Salman, O.H., Zaidan, A.A., Hashim, M.: Novel methodology for triage and prioritizing using "big data" patients with chronic heart diseases through telemedicine environmental. Int. J. Inf. Technol. Dec. Mak. **16**(5), 1211–1246 (2017)
6. Caicedo-Torres, W., García, G., Pinzón, H.: A machine learning model for triage in lean pediatric emergency departments. Wirel. Pers. Commun. **71**(2), 212–221 (2012)

7. Ghive, A.A., Patil, D.R.: Implementation of SVM and NB algorithms for classification of diseases and their treatments. Int. J. Adv. Sci. Eng. Technol. **3**(4), 141–145 (2015)
8. Lee, S., Weiss, E., Shumei, S.S.: Irregular breathing classification from multiple patient dataset using neural networks. IEEE Trans. Inf. Technol. Biomed. **16**(6), 1253–1264 (2012)

Author Index

Printed in the United States
By Bookmasters